Lecture Notes in Artificial Intelligence 833

Subseries of Lecture Notes in Computer Science
Edited by J. G. Carbonell and J. Siekmann

Lecture Notes in Computer Science

Edited by G. Goos and J. Hartmanis

Dimiter Driankov   Peter W. Eklund
Anca L. Ralescu (Eds.)

Fuzzy Logic and Fuzzy Control

IJCAI '91 Workshops on Fuzzy Logic
and Fuzzy Control
Sydney, Australia, August 24, 1991
Proceedings

Springer-Verlag
Berlin Heidelberg New York
London Paris Tokyo
Hong Kong Barcelona
Budapest

Series Editors

Jaime G. Carbonell
School of Computer Science, Carnegie Mellon University
Schenley Park, Pittsburgh, PA 15213-3890, USA

Jörg Siekmann
University of Saarland
German Research Center for Artificial Intelligence (DFKI)
Stuhlsatzenhausweg 3, D-66123 Saarbrücken, Germany

Volume Editors

Dimiter Driankov
Department of Computer and Information Science, Linköping University
S-58183 Linköping, Sweden

Peter W. Eklund
Department of Computer Science, University of Adelaide
GPO Box 498, Adelaide SA 5001, Australia

Anca L. Ralescu
Laboratory for International Fuzzy Engineering Resesarch
89-1 Yamashita-cho, Naka-ku, Yokohama, 231 Japan

CR Subject Classification (1991): I.2.3, I.5.1, J.6

ISBN 3-540-58279-7 Springer-Verlag Berlin Heidelberg New York
ISBN 0-387-58279-7 Springer-Verlag New York Berlin Heidelberg

CIP data applied for

Typesetting: Camera ready by author
SPIN: 10475354   45/3140-543210 - Printed on acid-free paper

Preface

Why have a fuzzy control/logic workshop at a major AI conference?

Despite the indisputable success of Fuzzy Control (FC) there remains an important issue which requires further investigation and more solid and thorough treatment. This issue, considered to be the weakest point of FC, is the lack of formally sound procedures for FC design and analysis based on fuzzy (or other types of quantitative/qualitative) process models. In other words, there is a strong need for the development of fuzzy dynamic systems theory with an emphasis on the modelling the linguistic structure of the process (or deriving it from existing quantitative/qualitative models). This theory should extend in a qualitative way the fundamental notions of state, controllability, and stability and provide a means for automatic derivation and analysis of fuzzy controllers allowing at the same time for easy incorporation of expert knowledge.

On the other hand, some recent developments in the area of qualitative modelling in AI seem to point in the direction of providing qualitative models for the behaviour of complex nonlinear systems using qualitative descriptions of phase-plane structures.

The ultimate goal is to develop a class of intelligent controllers that "understand" the phase-planes of complex nonlinear systems, sense the world, synthesize control commands, and affect processes. Accomplishing such a difficult task would be hard to imagine without the controller being able to understand and analyse the qualitative behaviour of a system, especially when the system is of higher order and operates in a nonlinear regime.

However, these two approaches to intelligent control, although aiming at similar objectives, have developed very much independently from each other and there has not been an exchange of ideas between the two scientific communities. In control theory terms, a fuzzy rule based formalism can be likened to a qualitative input/output model whereas, in AI the approach is akin to qualitative state-space description and performs the function of an internal representation of the process. Thus, the fuzzy control representation describes what an experienced process-operator does, rather than why he does it. The answer to the latter can only come from the internal representation of the process, i.e. its model.

It is in this context that the need for the workshop was conceived. Its purpose was seen in terms of providing a framework within which the similarities and differences between the two approaches could be highlighted and discussed.

The workshop programme and its goals

The morning session of the workshop (10.00-12.30 am) contained three long presentations (45 minutes each) – all three were given by Japanese participants and devoted to providing a complete picture of the Japanese large-scale effort in the area of fuzzy control and fuzzy expert systems. The latter are built using basic fuzzy control techniques augmented with object and frame-based knowledge representation techniques as well as some means for temporal and causal reasoning.

The speakers were able to pinpoint the need for fuzzy treatment in a number of comparatively large-scale applications, to systematize the lessons learned and specify in a systematic way the advantages and the limitations of the fuzzy approach as well as its integration with AI techniques. The talk by Professor Hirota [Deputy Director of the Laboratory for International Fuzzy Engineering (LIFE)], provided for an excellent description of major ongoing projects at LIFE, e.g. an expert system shell for automatic analysis of economic indicators incorporating quantitative, qualitative, and temporal information about these indicators; the linguistic control of a robot; and high-level conceptual analysis of images using fuzzy object and frame-based representations.

Professor Hirota [Hosei University, Tokyo] concentrated on explaining why the fuzzy approach has acquired such prominence in Japanese academic and industrial circles. He also gave a detailed picture of the very well orchestrated Japanese research in the area of fuzzy control and pointed out the benefits from close co-operation with industry.

Dr. Wakami [Matsushita Electric] gave an exhaustive survey of the applications of fuzzy control to a wide range of low-tech consumer products and also described Matsushita's research effort in the area of neuro-fuzzy controllers.

Thus the morning session achieved a number of goals, it gave;

- an introduction to the basic fuzzy techniques employed in fuzzy control and fuzzy expert systems, pointing out their advantages and limitations in the context of specific application areas;
- an idea how the fuzzy approach can be combined with basic AI techniques and approaches (frames, object-oriented approach, temporal reasoning, neural networks);
- a description of the applications domains where the fuzzy approach was successful and is worth applying.

The afternoon session was divided into three parts. The first part included two long-presentations (45 minutes each) concerning novel theoretical results about fuzzy control rules and the modeling of fuzzy dynamic systems.

Professor Prade, after summarising the main features of the existing fuzzy control rules and pointing out their limitations, proposed a new type of rule called a "gradual rule", e.g. "The more x has the property A, the more y has the property B" instead of the conventional interpretation, "If x is A then y is B". The main advantage of such a gradual rule is that the defuzzification part of a fuzzy control algorithm becomes unnecessary. The underlying motivation for getting rid of the latter is that whenever the controller stems from a single set of rules it seems that a preprocessing of these rules which transforms the fuzzy controller into a standard, possibly non-linear, control law is more reasonable than a two-step process that computes first a fuzzy control output and then defuzzifies it. However, the appropriateness of such control rules, though reasonable from a theoretical point of view, remains to be tested in real control applications.

Professor Pedrycz's presentation identified the major approaches to modeling dynamic fuzzy systems, concentrating on first-order systems. These approaches can be classified into three different types:

1. direct relational structures
2. state equations with a difference operator
3. referential fuzzy dynamical systems

Professor Pedrycz pointed out some similarities between system descriptions coming from AI-based qualitative reasoning and the referential fuzzy dynamic systems.

The second part of the afternoon session was devoted to short presentations (20 minutes each) describing work in the following areas:

– fuzzy systems modelling: Z. Cao reported a mapping algorithm for fuzzy reasoning as an alternative to the fuzzy rule-based reasoning. However, its relevance with respect to control applications remains to be investigated. B. Das reported work on dynamic systems based on fuzzy non-linear functions with emphasis on their temporal and uncertainty management aspects.
– design methods for fuzzy control: D. Stirling and S. Lamba described a fuzzy logic based expert system for stainless steel cold rolling mill. The approach taken incorporates a qualitative model of the process based on causal graphs and a set of heuristics based on IF-THEN fuzzy rules.
B. Graham described a method for the identification of a process-model from plant input/output data. The model is in the form of qualitative linguistic relationships, by adding on-line identification of the model he showed how a fuzzy adaptive controller can be constructed.
– applications of fuzzy control: Y. Darvish talked about a project to develop a fuzzy controller for the lateral axis of an autopilot for the Boeing 737. Kieronska and Vankatesh reported on the use of fuzzy control techniques in a robot's navigational system. Y. Katz identified the parts of a system for the management of large computer networks which require real-time fuzzy control. S. Libberstein talked about using fuzzy logic techniques in the construction of any-time algorithms.

The third part of the afternoon session was devoted to a discussion which concentrated on the following two topics:

– what should a fuzzy control design methodology consist of?
– the need for QM techniques in FC.

Concerning the first topic the following view emerged: It is important to distinguish between the following three tasks from which a methodology would consist:

- the elicitation of fuzzy control rules;
- the synthesis of a control law from the rules;
- the study of the properties of the control law.

The first requires careful analysis of what redundancy, completeness, and consistency mean for a set of fuzzy rules where subsets of rules can be fired in parallel. It has to be stressed that such an elicitation makes sense only if the process-operator's expertise is available. If it is lacking, a learning methodology based on neural nets might be more appropriate. However, in practice, it is likely that in most cases the expertise will be available, but without being sufficient for a precise determination of the controller, i.e. some learning will be required anyway.

The second task could be tackled by regression methods from fuzzy data. The third task is considered to be beyond the competence of knowledge-engineers and should be rather left to specialists in automatic control.

Thus, there is a strong need to develop methods and computerized tools for improving and automating each one of these three tasks in order to go beyond the existing fuzzy control development techniques.

One can see here that the proposed ingredients of the methodology rely heavily on the fact that there is an absence of a well-understood model of the process. In such cases one can resort to the process-operator's operating knowledge as a way of building the controller by implicitly taking into account his knowledge of the process.

But what about processes where the lack of a well-defined model is further aggravated by lack of expertise? The answer to this question, as all participants agreed, lies in the combination of AI-based qualitative reasoning and fuzzy control techniques.

The idea is to develop a system which by interacting with a qualitative description of the process, e.g. high-level symbolic description of a phase-plane, will automatically construct a fuzzy controller. More specifically, the system will derive the fuzzy control rules and the fuzzy sets describing the components of these rules which best fit certain desired behaviour resulting from the qualitative model of the process. This is especially attractive in cases when expertise does not yet exist. Furthermore, if any modifications are made to the physical system being modelled, they can be reflected only in the model, and the fuzzy rules and sets will be automatically adjusted as a result of these changes.

Another possibility is to create qualitative models based on fuzzy logic rather than existing AI techniques for qualitative modelling. In this case the relationships between the system-variables are defined in fuzzy terms, e.g. "slight increase in temperature will significantly increase the volume" and so on. This type of approach can be combined with such qualitative techniques as causal graphs and/or bond graphs which give good structural models and will be very well suited for representing analog, mechanical, and hydraulic systems.

At the end of the discussion J. Katz, the IJCAI official responsible for the IJCAI Workshop Programme, suggested the publication of workshop proceedings with full versions of the presented long and short papers. It was decided that work presented in the IJCAI Workshop on Fuzzy Logic should also be reflected in the proceedings.

This volume is a consequence of that initial suggestion. The editors would like to acknowledge the assistance of the Department of Computer and Information Science, The University of Linköping, the Department of Computer Science, The University of Adelaide, the Laboratory for International Fuzzy Engineering (LIFE), Seimens AG, SAAB Missiles AB, and an IJCAI '91 travel grant.

Dimiter Driankov
Peter W. Eklund
Anca Ralescu

Contents

Part I

Fuzzy Logic-based Reasoning and Control: Theoretical Aspects - Fuzzy Reasoning

Basic Issues on Fuzzy Rules and their Application to Fuzzy Control

Didier Dubois and Henri Prade[1]

Institut de Recherche en Informatique de Toulouse (IRIT), Université Paul Sabatier - CNRS, 118 route de Narbonne, 31062 Toulouse Cedex, France Tel. : (+33) 61.55.63.31 / 65.79 - Fax. : (+33) 61.55.62.39

Abstract. Fuzzy logic controllers have encountered an extraordinary success in a great variety of industrial applications in the last few years, especially in Japan. The principle of fuzzy controllers, first outlined by Zadeh[31] and then successfully experimented by Mamdani and Assilian[20], consists of synthesizing a control law for a system from fuzzy rules, usually provided by experts, which state the action(s) to do in typical situations, in contrast with the standard approach to automatic control which requires a model of the system to control. Each rule more or less applies to a fuzzy class of situations and an interpolation operation is performed between the conclusion parts of the selected rules, on the basis of the degrees of compatibility between the condition parts of these rules and the current situation encountered by the system. The reader is referred to Mamdani[19], Sugeno[24] for introductions and to Lee[16] and Berenji[5] for surveys. The basic methodology of fuzzy logic controllers was empirically developed in the late seventies and early eighties and has not changed much since. Recently, a revival of rather theoretically-oriented studies has been observed in order to build a strong methodology for fuzzy logic controllers. Thus the analytical comparison between a fuzzy controller and a proportional-integral controller[30], the limit behavior of fuzzy controllers[6], the stability of fuzzy controllers[27], [26], adaptive techniques for fuzzy controllers, e.g. [22]; [1]; [13], and the use of neural network methods for learning fuzzy rules and implementation issues[17] [28] have been discussed.
In this paper we concentrate our methodological study on issues related to the modelling of the set of fuzzy rules and the associated interpolation techniques and we point out several directions for further research.

1 Synthesizing a Control Law from Fuzzy Expert Rules

1.1 Graph view vs. functional view

The basic idea of fuzzy control approaches is to elaborate a control law from a set of fuzzy rules describing how an expert reacts to various situations in order to control the behaviour of a dynamic system. The problem is to represent a control law from incomplete and rough specifications. In the case of a precisely stated control law $y = f(x)$ where y is the command and x stands for a vector of observed variables, there is an equivalence between two views of the control law: (i) its functional description $y = f(x)$ and (ii) its graph description $\{(x, y) \in f\}$ where f also denotes the relation which holds between x and y if and only if $y = f(x)$. As pointed out in [9], these two views are no longer equivalent in the case of imprecise or fuzzy specifications and lead to two different ways of dealing with fuzzy rules. In fuzzy control we work with a collection of n rules of the form "if x is A_i then y is B_i" for $i = 1, \ldots, n$ (where x may still be a vector and A_i a subset on a Cartesian product of domains).

For simplicity let us first consider the case where A_i and B_i are ordinary (non fuzzy) subsets and $\cup_i A_i$ covers the domain of x. The graph view leads to consider the pairs (A_i, B_i) as part of a graph of a relation R such that;

$$\forall i, A_i \times B_i \subseteq R; \tag{1}$$

this leads to build the relation R_g that contains f, i.e. $\forall x, f(x) \in R_g(x) = \{y \mid x \; R_g \; y\}$,

$$R_g = \cup_{i=1,\ldots,n} A_i \times B_i, \tag{2}$$

The functional view leads us to view B_i as the image of A_i by a relation R, i.e.,

$$\exists R, \forall i, A_i \circ R \subseteq B_i, \tag{3}$$

which is equivalent to

$$R \subseteq A_i \to B_i = \overline{A_i} \cup B_i, \tag{4}$$

where the union notation $\cup$ is also used to denote the Cartesian co-product $A \cup B = \overline{\overline{A} \times \overline{B}}$. This leads to build the relation R_f,

$$R_f = \cap_{i=1,\ldots,n} (\overline{A_i} \cup B_i) \tag{5}$$

But, as it can be seen on Figs 1 and 2, we generally have,

$$R_f \subset R_g, \tag{6}$$

if $\bigcup_{i=1,\ldots,n} A_i = X$, the range of x, i.e. the description provided by R_f is less imprecise than the one given by R_g. Note that R_g can be defined using the inclusion,

$$\forall i, A_i \circ \overline{R} \subseteq \overline{B_i}, \tag{7}$$

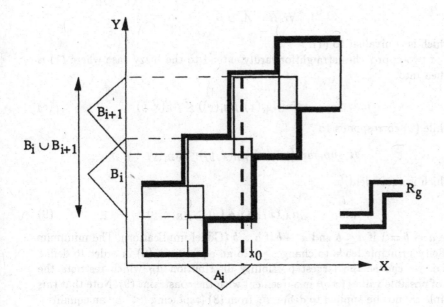

Fig. 1: The graph view

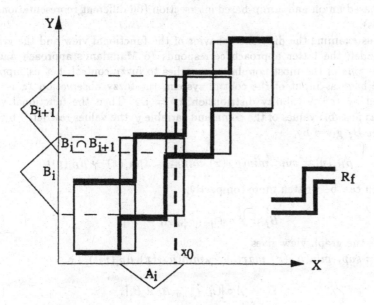

Fig. 2: The functional view

which states that $\overline{B_i}$ contains the elements not related to elements in A_i. It leads to,

$$\forall i, \overline{R} \subseteq \overline{A_i} \cup \overline{B_i}$$

which is equivalent to (1).

The two approaches straightforwardly extend to the fuzzy case where (1) is modified into,

$$max_{i=1,\ldots,n} min(\mu_{A_i}(x), \mu_{B_i}(y)) \leq \mu_R(x, y) \tag{8}$$

while (3) corresponds to

$$\forall i, sup_x min(\mu_{A_i}(x), \mu_R(x, y)) \leq \mu_{B_i}(y)$$

which is equivalent to

$$min_{i=1,\ldots,n} \mu_{A_i}(x) \to \mu_{B_i}(y) \geq \mu_R(x, y) \tag{9}$$

where $a \to b = 1$ if $a \leq b$ and $a \to b$ if $a > b$ (Gödel implication). The minimum specificity principle leads to change $\geq$ into an equality in (9) in order to define R_f, i.e. we choose the largest possibility distribution μ_R which restricts the pairs of possible values (x, y) in agreement with the constraint (9). Note that this principle cannot be applied to define R_g from (8) (replacing "$\leq$" by an equality), since (8) expresses a lower bound for μ_R, but it can be applied to define $\overline{R_g}$ from (7). As it can be observed, the way the representations of the rules are aggregated in the graph view and in the functional view differ[3]; we respectively perform a max-based union and a min-based intersection (on different representations of the rules).

Let us examine the different behavior of the functional view and the graph view model; the latter approach corresponds to Mamdani's approach and is presently one of the most popular approaches to fuzzy control. Let us suppose that we have as input of the control system, the fuzzy observation "x is A", modelled by the possibility distribution $\pi_x = \mu_A$. Then the functional view yields, as possible values of the command variable y, the values restricted by the fuzzy set B_f given by,

$$\mu_{B_f}(y) = sup_x \ min(\mu_A(x), min_{i=1,\ldots,n} \mu_{A_i}(x) \to \mu_{B_i}(y)),$$

which can be written more compactly,

$$B_f = A \circ [\cap_{i=1,\ldots,n} A_i \to B_i], \tag{10}$$

while the graph view gives
$$\mu_{B_g}(y) = sup_x \ min(\mu_A(x), max_{i=1,\ldots,n} min(\mu_{A_i}(x), \mu_{B_i}(y))), \text{ i.e.,}$$

$$B_g = A \circ [\cup_{i=1,\ldots,n} A_i \times B_i]. \tag{11}$$

When $A = \{x_0\}$, denotes $\alpha_i = \mu_{A_i}(x_0)$, we get,

$$B_f = \cap_{i=1,\ldots,n} \alpha_i \to B_i \text{ (with } \mu_{\alpha_i \to B_i}(y) = \alpha_i \to \mu_{B_i}(y)),$$

and

$$B_g = \cup_{i=1,\ldots,n} \; \alpha_i \times B_i \; (\text{with } \mu_{\alpha_i \times B_i}(y) = min(\alpha_i, \mu_{B_i}(y))).$$

Another often noticed difference between the functional view and the graph view is the impossibility, with the first approach, to reason rule-by-rule in presence of imprecision $(A \neq \{x_0\})$. Indeed,

$$A \circ [\cap_{i=1,\ldots,n} \; A_i \to B_i] \subset \cap_{i=1,\ldots,n} \; A \circ (A_i \to B_i)$$

For instance, if $A = A_j \cup A_k$ then $A \circ [\cup_{i=1,\ldots,n} A_i \to B_i] \subseteq B_j \cup B_k$, while $(A_j \cup A_k) \circ (A_j \to B_j) = V$ generally, where V is the whole domain of y. By contrast $A \circ [\cup_{i=1,\ldots,n} \; A_i \times B_i] = \cup_{i=1,\ldots,n} \; A \circ (A_i \times B_i)$.

1.2 Limitations of the graph view

As it can be seen in Figs 1 and 2, in the case where the A_i and B_i are non-fuzzy, for an input $x = x_0$, which is known to belong to $A_i \cap A_j$, the functional view will lead to the conclusion that $y \in B_i \cap B_j$, as expected, while the graph view yields the more imprecise conclusion $y \in B_i \cup B_j$. Indeed (8) can be written in such case as,

$$max(min(\mu_{A_i}(x_0), \mu_{B_i}(y)), min(\mu_{A_j}(x_0), \mu_{B_j}(y)) \leq \mu_R(x_0, y),$$

or equivalently if $\mu_{A_i}(x_0) = \mu_{A_j}(x_0) = 1$,

$$max(\mu_{B_i}(y), \mu_{B_j}(y)) = \mu_{B_i \cup B_j}(y) \leq \mu_R(x_0, y),$$

where $\mu_R(x_0, y)$ restricts the values of y associated to $x = x_0$. By contrast (9) gives,

$$min(\mu_{A_i}(x_0) \to \mu_{B_i}(y), \mu_{A_j}(x_0) \to \mu_{B_j}(y)) = \mu_{B_i \cap B_j}(y) \geq \mu_R(x_0, y),$$

since $1 \to a = a$.

Moreover, as pointed out by Di Nola *et al.*[7], if we have $A = A_i$, and A_i is a fuzzy set which intersects other A_j, then generally the response B_g, obtained by (11) has a wider support than B_i. In other words, it does not coincide with usual deduction. This is not the case with the functional view for which we have $A_i \circ [\cap_{j=1,\ldots,n} A_j \to B_j] \subseteq B_i$. The differences between the graph view and the functional view, point out the need for a better understanding of the meaning of fuzzy rules.

2 Various Kinds of Fuzzy Rules

As recently discussed in Dubois and Prade[10, 11], we can distinguish between

- gradual rules of the form "the more x is A, the more y is B"
- certainty rules of the form "the more x is A, the more *certain* y is B"
- possibility rules of the form "the more x is A, the more *possible* y is B"

Fuzzy relations R modelling gradual rules obey equations of the form $A \circ R \subseteq B$, i.e.,

$$\forall x, min(\mu_A(x), \mu_R(x, y)) \leq \mu_B(y), \tag{12}$$

since the greater $\mu_A(x)$, and the more y is in relation with x, the greater $\mu_B(y)$ should be. This leads to $R = A \rightarrow B$ where $\rightarrow$ is Gödel implication, i.e. we recover the rules encountered in the functional view. Another type of gradual rule obeys the dual inequality $A \circ \overline{R} \subseteq \overline{B}$ which leads to $max(1 - \mu_A(x), \mu_R(x, y)) \geq \mu_B(y)$ and $\mu_R(x, y) \geq \mu_A(x) \ \& \ \mu_B(y)$ where "$\&$" is a non-commutative conjunction such that $a \ \& \ b = b$ if $a + b > 1$ and 0 otherwise. They correspond to gradual rules expressing that "the more x is A and the less y is an image of x, then the less y is B". We may as well write $R = \overline{(A \rightarrow \overline{B})}$.

If we are only interested in the least specific crisp solution to (12), we get a crisp relation R_0 such that $\mu_{R_0}(x, y) = 1$ if $\mu_A(x) \leq \mu_B(y)$ and 0 otherwise. The main difference in behavior between the crisp and the fuzzy representations of this kind of rule is that the fuzzy representation obeys *modus ponens* ($A' \subseteq A \Rightarrow A' \circ (A \rightarrow B) = B$) while with the crisp solution, if $A' \subset A$, we may observe $A' \circ R_0 \subset B$ (strict inclusion).

Certainty rules correspond to the following kind of equations,

$$\forall x, \ \mu_A(x) \leq C_x(B), \tag{13}$$

where $C_x(B)$ is a degree of certainty of B which depends on x, of the form

$$C_x(B) = \inf_y \pi^x(y) \Rightarrow \mu_B(y). \tag{14}$$

The idea is to interpret $\mu_A(x)$ as the degree of certainty of B, and to determine π^x (defined on Y for a given x), as the least specific solution of (13).

C_x should be such that the following properties hold,

if $\mu_A(x) = 1$ then $\pi^x = \mu_B$ (since B is sure)
if $\mu_A(x) = 0$ then $\forall y, \ \pi^x(y) = 1$ (since B is totally uncertain)
if $\mu_A(x) = \alpha < 1$ then $\forall y, \ \pi^x(y) \geq 1 - \alpha$ (using the duality between the
                                              degree of certainty of B and
                                              degree of possibility of "not B").

In other words, a reasonable model for these rules is to let the equivalent fuzzy relation R satisfy,

$$\mu_R(x, y) = \pi^x(y) = max(\mu_B(y), 1 - \mu_A(x)), \tag{15}$$

i.e. $R = \overline{A \times \overline{B}}$, which corresponds to Dienes' implication. Implication "$\Rightarrow$" defining C_x in (14) turns out to be the reciprocal of Gödel implication, i.e. $r \Rightarrow s = (1 - s) \rightarrow (1 - r)$.

Possibility rules correspond to the following kind of equation:

$$\forall x, \; \mu_A(x) \leq \Delta_x(B), \tag{16}$$

where $\Delta_x(B)$ is the degree to which each element in the support of B is possible. $\Delta_x(B)$ is not a degree of possibility $\Pi_x(B)$ in the sense of [32]. Indeed, $\Pi_x(B)$ is rather a degree of consistency with the available evidence while $\Delta_x(B)$ corresponds to a stronger possibility concept. This is defined by $\Delta_x(B) = \inf_{y \in B} \pi^x(y)$ when B is non-fuzzy; see [11] for the general case). Namely (16) should be understood as follows;

if $\mu_A(x) = 1$ then $\pi^x(y) \geq \mu_B(y)$, i.e. all y in support of B have possibility
at least equal to $\mu_B(y)$
if $\mu_A(x) = 0$ then no constraints exists on $\pi^x(y), \forall y$
if $\mu_A(x) = \alpha$ then $\pi^x(y) \geq min(\mu_B(y), \alpha)$.

This leads to define the fuzzy relation R such that $\pi^x(y) = \mu_R(x, y)$ as follows:

$$\mu_R(x, y) \geq min(\mu_A(x), \mu_B(y)). \tag{17}$$

This corresponds to the graph view model of fuzzy rules, as proposed by Mamdani. The response of a system of parallel fuzzy possibility rules becomes more natural. Namely from the two crisp rules

if $x \in A_1$ then $x \in B_1$ is possible
if $x \in A_2$ then $x \in B_2$ is possible

and from the fact $x \in A_1 \cap A_2$, one concludes that $x \in B_1$ is possible as well as $x \in B_2$, hence x may lie in B_1 or B_2. The response $x \in B_1$ or B_2 is thus natural. Note that possibility rules may be useful in analogical reasoning in order to express rules like "the more x and x' are similar, the more possible the similarity of y and y'".

3   Fuzzy Rules for Representing Control Laws

If we leave the defuzzification problem aside (see Yager [29]), Filev and Yager[12] on this point), it is clear that gradual rules sound more adapted to the specification of a control law than the other types of rules. Indeed, a certainty rule creates a level of uncertainty on its conclusion whenever the input x gets away from the core of the condition. This is not very useful for characterizing the image of an input x through a function. Similarly, possibility rules correspond to a form of reasoning by analogy, whereby the existence of pairs (a, b_1) and (a, b_2) in the description of f leads us to consider that b_1, as well as b_2, can be the image of a.

Describing a control law is more a matter of interpolation. Namely the most elementary way of partially describing f is to supply a finite subset of its graph $\{(a_i, b_i), i = 1, \ldots, n\}$ where $(a_i, b_i) \in f$, $a_1 < a_2 < \ldots < a_n$. The crudest way of completing the graph is by linear interpolation:

if $a_i \leq x \leq a_{i+1}$ then $f(x) = \frac{b_{i+1}(x-a_i)+b_i(a_{i+1}-x)}{a_{i+1}-a_i}$.

This proposal is questionable. A more cautious way of representing f is to consider that $\forall a_i, a_{i+1}$, the middle point of the interval $[a_i, a_{i+1}]$ is where imprecision on $f(x)$ is maximal. When $a_i < a_{i+1} \Rightarrow b_i < b_{i+1}$ (monotonicity in the data), assuming the monotonicity of f sounds reasonable, we may conclude that $f(\frac{a_i+a_{i+1}}{2}) \in [b_i, b_{i+1}]$. Accepting linear interpolation to complete the partial description of the obtained relation R, i.e. to link the pairs $\{(a_i, b_i), i = 1, \ldots n\} \cup \left\{ \left(\frac{(a_i+a_{i+1})}{2}, [b_i, b_{i+1}] \right), i = 1, \ldots, n \right\}$, we obtain the representation of R by diamonds, as in Fig. 3, which presupposes that the slope of f is only slowly evolving.

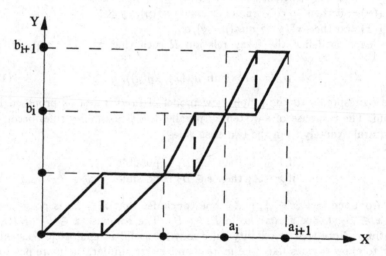

Fig. 3 : A cautious interpolation

Another way of interpreting $\{(a_i, b_i), i = 1, \ldots, n\}$, is to exploit gradual rules of the form "the closer x is to a_i, the closer y is to b_i". The first problem is to represent "close to a_i", by means of a fuzzy set A_i. It seems natural to assume that $\mu_{A_i}(a_{i-1}) = \mu_{A_i}(a_{i+1}) = 0$ since there are special rules adapted to the cases $x = a_{i-1}, x = a_{i+1}$. Moreover, if $x \neq a_i, \mu_{A_i}(x) < 1$ for $x \in (a_{i-1}, a_{i+1})$, since information is only available for $x = a_i$. Hence A_i should be a fuzzy interval with support (a_{i-1}, a_{i+1}) and core $\{a_i\}$.

In addition, by symmetry, since the closer x is to a_{i-1}, the farther it is from $a_i, \mu_{A_{i-1}}(x)$ would decrease when $\mu_{A_i}(x)$ increases, and $\mu_{A_i}(\frac{a_i+a_{i+1}}{2}) = \mu_{A_{i-1}}(\frac{a_{i-1}+a_i}{2}) = 0.5$. The simplest way of achieving this is to let,

$$\forall x \in [a_{i-1}, a_i], \ \mu_{A_{i-1}}(x) + \mu_{A_i}(x) = 1.$$

Clearly, the conclusion parts of the rules should involve fuzzy sets B_i whose meaning is "close to b_i", with similar convention (see Fig. 4). In other words, each pair (a_i, b_i) is understood as "the more x is A_i, the more y is B_i", i.e. "the closer x is to a_i, the closer y is to b_i". In that case the output to $x = a$, where $a_{i-1} < a < a_i$ is,

$$B = (\alpha_{i-1} \to B_{i-1}) \cap (\alpha_i \to B_i),$$

with $\alpha_{i-1} + \alpha_i = 1, \alpha_{i-1} = \mu_{A_{i-1}}(a), \alpha_i = \mu_{A_i}(a)$. Now it is easy to verify, using Gödel's implication, that,

$$\mu_B(y) = 1 \text{ if } \mu_{B_{i-1}}(y) \geq \alpha_{i-1} \text{ and } \mu_{B_i}(y) \geq \alpha_i.$$

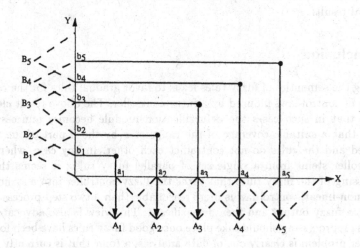

Fig. 4: The closer x is to a_i, the closer y is to b_i

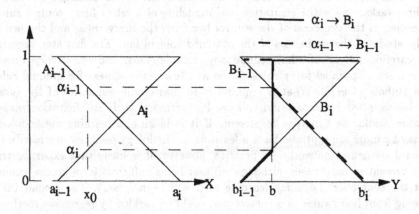

Fig. 5: The core of $(\alpha_i \to B_i) \cap (\alpha_{i-1} \to B_{i-1})$ is the result of a linear interpolation.

But since $\mu_{B_{i-1}}(y) + \mu_{B_i}(y) = \alpha_{i-1} + \alpha_i = 1$, there is a single value $y = b$ such that $\mu_B(b) = 1$ and it is such that $b = \mu_{B_i}^{-1}(\alpha_i) = \mu_{B_{i-1}}^{-1}(\alpha_{i-1})$ on $[b_{i-1}, b_i]$. Particularly, if $A_{i-1}, A_i, B_i, B_{i-1}$ have triangular shapes, the core b of B is exactly the result of the linear interpolation, i.e. $b = \mu_{B_i}^{-1}(\alpha_i) \cdot b_{i-1} + a_i \cdot b_i$ (see Fig. 5); this corresponds to Sugeno and Nishida's[24] inference method where the conclusion parts of the rules are precise values b_i and where a linear interpolation is performed on the basis of the degrees of matching $\alpha_i = \mu_{A_i}(a)$. Hence gradual rules do model interpolation, linear interpolation being retrieved as a particular case.

More generally, the interpolation problem where only fuzzy points that do not overlap are given (rather than precise pairs (a_i, b_i)), is a topic for further research; however, see the proposal by Koczy et al.[14]. See Lowen[18] for a theoretical result.

4 Conclusion

Analyzing the semantics of fuzzy rules leads to favor gradual rules for the representation of control laws induced by a fuzzy controller. The above result clearly indicates that in such cases the defuzzification module becomes unnecessary, provided that a suitable coverage of the range of x by the if-part of the rules is adopted and the rules do not contradict each other. In any case, whenever the controller stems from a single set of parallel fuzzy rules, it seems that a preprocessing of the fuzzy rules that turns the fuzzy controller into a standard, possibly non-linear, control law is more reasonable than a two step process that computes a fuzzy output and then defuzzifies it. This view is also advocated in [15]. This preprocessing should take place once good fuzzy rules have been found. This latter problem is clearly one of data analysis, a topic that is currently very active in the fuzzy set community[4].

It is important, in the fuzzy control methodology, to distinguish between three tasks: one is the elicitation and modeling of a set of fuzzy control rules, another is the synthesis of the control law from the fuzzy rules, and the last is the study of the properties of the obtained control law. The first step requires a careful analysis of what redundancy, completeness, and consistency means for a set of parallel fuzzy rules. Some results already appear for gradual rules in Dubois et al.[8]. What is required is to master the semantics of the fuzzy rules supplied by expert individuals. Furthermore, such an elicitation makes sense insofar as expertise is present. If it is lacking, a learning methodology may be more appropriate. Such a learning methodology need not necessarily be based on neural networks. In practice, however, it is likely that expertise will be present in most cases, although without being sufficient for a precise tuning of the controller, i.e. some learning step will be necessary. The second task, going from fuzzy rules to a control law, could be tackled by regression methods from fuzzy data in case of an incomplete set of rules. The third task is beyond the competence of knowledge engineers, and should be rather left to automatic control specialists.

On the whole, we feel the necessity to develop methods and computerized tools for improving the state of the art on each of these three tasks, in order to go beyond the original fuzzy logic control methodology.

In the long term, it would be interesting to relate the fuzzy knowledge available about the control of a (complex) system and the modelling of this system by means of fuzzy qualitative equations. In that respect, works on fuzzy differential equations[23];[2] are particularly worth considering. Another issue would be to look for relations which might exist between fuzzy control techniques and robust control concerns in classical automatic control, e.g. [21].

References

1. Andersen T.R., Nielsen, S.B: An efficient single output fuzzy control algorithm for adaptive applications. Automatica 21, (1985) 539-545.
2. Aubin J.P.: Mathematical Methods of Artificial Intelligence. Lab. CEREMADE Université de Paris-Dauphine (1991).
3. Baldwin J.F., Guild, N.C.F.: Modelling controllers using fuzzy relations. Kybernetes 9 (1980) 223-229.
4. Bandemer H.: From fuzzy data to functional relationships. Mathematical Modelling 9(6)(1987) 419-426.
5. Berenji H.R.: Fuzzy logic controllers. in: An Introduction to Fuzzy Logic Applications in Intelligent Systems. (R.R. Yager, L.A. Zadeh, eds.), Kluwer Academic, 1991.
6. Buckley J.J.: Fuzzy controller: further limit theorems for linear control rules. Fuzzy Sets and Systems 36 (1990) 225-233.
7. Di Nola A., W. Pedrycz and S. Sessa: An aspect of discrepancy in the implementation of modus ponens in the presence of fuzzy quantities, International Journal of Approximate Reasoning (3) 259-265, (1989).
8. Dubois D., Martin-Clouaire, R., Prade, H.: Practical computing in fuzzy logic. in: Fuzzy Computing - Theory, Hardware, and Applications, (M.M. Gupta, T. Yamakawa, eds.), North-Holland, Amsterdam, 11-34, (1988).
9. Dubois D. and H. Prade: A typology of fuzzy "if- then-rules. Proceedings of the International Fuzzy Systems Association (IFSA) Congress, Seattle, Wash., Aug. 6-11, (1989) 782-785.
10. Dubois D., Prade, H.: Fuzzy sets in approximate reasoning - Part 1: Inference with possibility distributions. Fuzzy Sets and Systems 40 (1991) 143-202.
11. Dubois D., Prade, H.: Fuzzy rules in knowledge-based systems - Modelling gradedness, uncertainty and preference. in: An Introduction to Fuzzy Logic Applications in Intelligent Systems (R.R. Yager, L.A. Zadeh, eds.), Kluwer Academic, (1992).
12. Filev D.P., Yager, D.R.: A generalized defuzzification method via bad distributions. International Journal of Intelligent Systems 6(7) (1991) 687-697.
13. Guely F. and I. Todo.: A new fuzzy self-organizing control algorithm and its application to the control of a robot arm. Proceedings of the 20th International Symposium on Industrial Robots (ISIR), Tokyo, (1989), 137-144.
14. Koczy L.T., Hirota K., Juhasz A.: Interpolation of 2 and $2k$ rules in fuzzy reasoning. Proceedings of the International Fuzzy Engineering Symposium '91 (IFES'91), Yokohama, Japan, (1991) 206-217.

15. Kuipers B., Astrom, K.: The composition of heterogenous control laws. Report AI90-138, Artificial Intelligence Laboratory, University of Texas, Austin, Tex.. Revised version in Proceedings 1991 American Control Conference (1990).

16. Lee C.C.: Fuzzy logic in control systems: fuzzy logic controller - Parts 1&2. IEEE Trans. on Systems, Man and Cybernetics 20(2) (1990) 404-435.

17. Lee C.C.: A self-learning rule-based controller employing approximate reasoning and neural net concepts. International Journal of Intelligent Systems 6 (1991) 71-93.

18. Lowen R.: A fuzzy Lagrange interpolation theorem. Fuzzy Sets and Systems 34 (1990) 33-38.

19. Mamdani E.H.: Application of fuzzy logic to approximate reasoning using linguistic systems. IEEE Trans. Comput. 26 (1977) 1182-1191.

20. Mamdani E.H., Assilian, S.: An experiment in linguistic synthesis with a fuzzy logic controller. International Journal of Man-Machine Studies 7 (1975) 1-13.

21. Petersen I.R.: A new extension to Kharitonov's theorem. IEEE Trans. on Automatic Control 35(7) (1990) 825-828.

22. Procyk T.J. Mamdani,E.H.: A linguistic self-organizing controller. Automatica 15 (1979) 15-30.

23. Shaw W.T., Grindrod, P.: Investigation of the potential of fuzzy sets and related approaches for treating uncertainties in radionuclide transfer predictions. Final Rep. CEC contract ETCC-0003, Environmental Sciences Department, INTEGRA-ECL, Henley-on-Thames, U.K., (1989)

24. Sugeno M.: An introductory survey of fuzzy control. Information Science 36 (1985) 59-83.

25. Sugeno M. Nishida, M.: ' Fuzzy control of model car. Fuzzy Sets and Systems 16 (1985) 103-113.

26. Tanaka K., M. Sugeno, M.: Stability analysis and design of fuzzy control systems. Fuzzy Sets and Systems, (1991).

27. Wang P.Z., H.M. Zhang, H.M, Xu, W.: PAD-analysis of fuzzy control stability. Fuzzy Sets and Systems 38 (1990) 27-42.

28. Yager R.R.: Implementing fuzzy logic controllers using a neural network framework. Tech. Report #MII-1005, Machine Intelligence Institute, Iona College, New Rochelle, N.Y., (1990).

29. Yager R.R.: An alternative procedure for the calculation of fuzzy logic controller values. Tech. Report #MII-1014, Machine Intelligence Institute, Iona College, New Rochelle, N.Y., (1990)

30. Ying H., Siler, W. , Buckley J.J.: Fuzzy control theory: a nonlinear case. Automatica 26(3) (1990) 513-520.

31. Zadeh L.A.: Outline of a new approach to the analysis of complex systems and decision processes. IEEE Trans. on Systems, Man and Cybernetics 3:28-44, (1973).

32. Zadeh L.A.: Fuzzy sets as a basis for a theory of possibility. Fuzzy Sets and Systems 1:3-28, (1978).

Algebraic Structures of Truth Values in Fuzzy Logic

Masao Mukaidono

Department of Computer Science, Meiji University, 1-1-1 Higashi-mita, Tama-ku,
Kawasaki-shi, 214 Japan, Tel:+81-44-934-7450, Fax:+81-44-934-7912,
E-mail:mukaido@jpnmu11.bitnet

Abstract. It is well known that in two-valued logic the set of truth values 0 and 1 with the logical operations AND($\land$), OR($\lor$) and NOT($\lnot$) satisfy the axioms of Boolean algebra. The characteristics of Boolean algebra provide the fundamental properties of AI systems if they are developed based on the classical two-valued logic; for example, they can not treat ambiguous states which are frequent in the real world. Fuzzy logic was introduced to treat such ambiguous (or not certainly definable) propositions of daily life. Up until now in fuzzy logic different truth values have been used. For example, the typical truth values used in fuzzy logic are numeric values on the unit interval $[0, 1]$. Another example of truth values used in fuzzy logic are interval truth values. These are pairs of numeric truth values (a, b), $a, b \in [0, 1]$, where a and b mean necessity and possibility degrees of the truth value, respectively. On the other hand, L. A. Zadeh introduced fuzzy truth values, which are fuzzy sets on $[0, 1]$. Of course, two-valued truth values 0 and 1 are special cases of numerical truth values, and numerical truth values are special cases of interval truth values. Furthermore, interval truth values are special cases of linguistic truth values[2].

In this paper, algebraic structures of numerical truth values, interval truth values and fuzzy truth values, are examined when three kinds of logic operations AND($\land$),OR($\lor$) and NOT($\lnot$) are defined on these truth values and their fundamental properties are investigated. It will be very important to know the algebraic structures of truth values in fuzzy logic if fuzzy logic is to be used in AI systems to treat ambiguity, since the framework for treating degrees of truth or falsity is determined fundamentally by the algebraic structure of the truth values used in the fuzzy logic.

1 Truth Values in Two-Valued Logic

Let B be a set of 0 and 1, that is, $B = \{0, 1\}$, the set of truth values of two-valued logic. Let a and b be any elements of B. Then, three logic operations AND($\land$), OR($\lor$) and NOT($\lnot$) are defined on B as follows:

$$a \land b \begin{cases} 0 \text{ if } a = 0 \text{ or } b = 0 \\ 1 \text{ otherwise} \end{cases} \tag{1}$$

$$a \vee b \begin{cases} 0 \text{ if } a = 0 \text{ or } b = 0 \\ 1 \text{ otherwise} \end{cases} \tag{2}$$

$$\neg a \begin{cases} 1 \text{ if } a = 0 \\ 0 \text{ otherwise} \end{cases} \text{ where } a, b \in B \tag{3}$$

It is well known that the algebraic system $\langle B, \wedge, \vee, \neg \rangle$, i.e. the set B and logic operations $\wedge, \vee, \neg$ defined on B, satisfies the following laws:

1. Commutative laws: $a \wedge b = b \wedge a, \quad a \vee b = b \vee a$,
2. Idempotent laws: $a \wedge a = a, \quad a \vee a = a$,
3. Absorption laws: $a \wedge (a \vee b) = a, \quad a \vee (a \wedge b) = a$,
4. Associative laws: $a \wedge (b \wedge c) = (a \wedge b) \wedge c, \quad a \vee (b \vee c) = (a \vee b) \vee c$,
5. Distributive laws: $a \wedge (b \vee c) = (a \wedge b) \vee (a \wedge c), \quad a \vee (b \wedge c) = (a \vee b) \wedge (a \vee c)$,
6. Double negations laws: $\neg \neg a = a$,
7. Least element: $0 \vee a = a, \quad 0 \wedge a = 0$,
8. Greatest element: $1 \vee a = 1, \quad 1 \wedge a = a$,
9. de Morgan's laws: $\neg(a \wedge b) = \neg a \vee \neg b, \quad \neg(a \vee b) = \neg a \wedge \neg b$,
10. Complementary laws: $a \wedge \neg a = 0, \quad a \vee \neg a = 1$, where $a, b, c \in B$

The algebraic system satisfying 1.-10. is called a <u>Boolean algebra</u>. Truth values of two-valued logic satisfy the axioms of a Boolean algebra, that is it is well known that the algebraic system $\langle B, \wedge, \vee, \neg \rangle$ is a Boolean algebra.

2 Numerical Truth Values

A numerical truth values in fuzzy logic is defined to be an element of the closed interval $[0, 1]$. as follows.

Definition 1: a is a <u>numerical truth value</u> if and only if a is an element of $[0, 1]$.
Let V be the set of all numerical truth values, that is $V = [0, 1]$.
Let a, b, c be numerical truth values and logic operations $\text{AND}(\wedge), \text{OR}(\vee)$ and $\text{NOT}(\neg)$ are defined as follows:

$$a \wedge b = min(a, b) \tag{4}$$

$$a \vee b = max(a, b) \tag{5}$$

$$\neg a = 1 - a \quad \text{where } a, b \in V = [0, 1]. \tag{6}$$

The definitions of (1),(2) and (3) are special cases of the above definitions (4), (5) and (6), respectively. It is easily shown that the algebraic system $\langle B, \wedge, \vee, \neg \rangle$ satisfies the above laws 1.-9. But it does not satisfy 10., i.e. Complementary laws, as follows.

Lemma 1: Numerical truth values do no satisfy the complementary laws.
Proof 1: For example, $a \wedge \neg a = 0.2 \neq 0$ and $a \vee \neg a = 0.8 \neq 1$ if $a = 0.2$. This is a counter example showing that numerical truth values do not satisfy the complementary laws.

So, the algebraic system $\langle B, \wedge, \vee, \neg \rangle$ is not a Boolean algebra. But, instead of 10., Complementary laws, the algebraic system $\langle B, \wedge, \vee, \neg \rangle$ satisfies the following laws called <u>10. Kleene's laws</u>:

$$(a \vee \neg a) \wedge (b \wedge \neg b) = b \wedge \neg b, \quad (a \vee \neg a) \vee (b \wedge \neg b) = a \vee \neg a.$$

Kleene's laws are equivalent to the condition $a \vee \neg a \geq b \wedge \neg b$.

Lemma 2: Numerical truth values satisfy Kleene's laws.

Proof 2: $a \vee \neg a = max(a, 1 - a) \geq 1/2$ for any a in $[0,1]$, and $b \wedge \neg b = min(b, 1 - b) \leq 1/2$ for any b in $[0, 1]$ (Fig. 1 (a),(b)). Therefore, for any a and b in $[0, 1]$, $a \vee \neg a \geq 1/2 \geq b \wedge \neg b$, i.e. the set of all numerical truth values satisfies Kleene's laws.

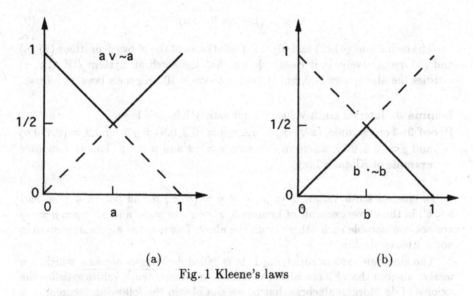

(a)   (b)
Fig. 1 Kleene's laws

If the complementary laws are satisfied, then Kleene's laws are satisfied, because $a \vee \neg a = 1 \geq b \wedge \neg b = 0$. But the converse is not true, i.e. Kleene's laws are weaker conditions than the complementary laws. The algebraic system satisfying 1-9. and 10. is called a Kleene algebra. That is, numerical truth values satisfy the axioms of Kleene algebra .

Theorem 1: The algebraic system $\langle B, \wedge, \vee, \neg \rangle$ is a Kleene algebra.

Proof 1: It is evident from the above discussions.

3 Interval Truth Values

A pair of two numerical truth values (a, b) such as $a \leq b$ is defined to be an interval truth value, as follows.

Definition 2: $x = (a, b)$ is an interval truth value if and only if $a, b \in V = [0, 1]$ and $a \leq b$ (fig. 2 (b)). Let W be the set of all interval truth values, that is

$$W = \{(a, b)| \ a \leq b, \ a, b \in [0, 1]\}$$

For interval truth values, three logic operations AND($\wedge$), OR($\vee$) and NOT($\neg$) are defined as follows.

Let $x = (xn, xp), y = (yn, yp)$ be two interval truth values. Then,

$$x \wedge y = (min(xn, yn), min(xp, yp)) \tag{7}$$

$$x \vee y = (max(xn, yn), max(xp, yp)) \tag{8}$$

$$\neg x = (1 - xp, 1 - xn). \tag{9}$$

The definitions (4),(5) and (6) are special cases of the above definitions (7),(8) and (9), respectively. It is easely shown that the algebraic system $\langle W, \wedge, \vee, \neg \rangle$ satisfies the above laws 1- 9. but it does not satisfy 10., Kleene's laws, as follows.

Lemma 3: Interval truth values do not satisfy Kleene's laws.
Proof 3: For example, $(x \vee \neg x) \wedge (y \wedge \neg y) = (0.3, 0.6) \neq y \wedge \neg y$ if $x = (0.4, 0.6)$ and $y = (0.3, 0.7)$, where in this case $x = \neg x$ and $y = \neg y$. This is a counter example of Kleene's laws.

In interval truth values, $x = (a, b) \leq y = (a', b')$ if and only if $a \leq a'$ and $b \leq b'$. In the above example of Lemma 3, $x \vee \neg x = x = \neg x$ and $y \wedge \neg y = y = \neg y$ are not comparable each other. From the above Lemma, the algebraic system is not a Kleene algebra.

The algebraic system satisfying 1- 9. is called de Morgan algebra, which is a weaker algebra than Kleene algebra. The set of interval truth values satisfies the axioms of de Morgan algebra, that is, we can obtain the following theorem.

Theorem 2: The algebraic system $\langle W, \wedge, \vee, \neg \rangle$ is a de Morgan algebra.
Proof 2: It is evident from the above discussions.

The interval truth values were extended by the authors [2] to such truth values as,

$$W = \{(a, b)| \ a, b \in [0, 1]\},$$

where the condition $a \leq b$ was ommitted. For the extended interval truth values, we can clarify many interesting properties under the same logic operations of (7),(8) and (9) and it was shown that it is also a de Morgan algebra [2].

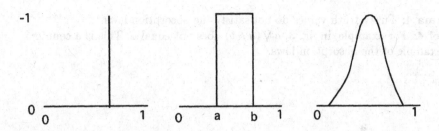

(a) Numerical Truth Values (b) Interval Truth Values (c) Fuzzy Truth Values
Fig. 2 Truth values in fuzzy logic

4 Fuzzy Truth Values

A fuzzy truth value is defined to be a fuzzy set on the closed interval $V = [0, 1]$, as follows.

Definition 3: a is a <u>fuzzy truth value</u> if and only if a is a fuzzy set on $[0, 1]$.
Let L be the set of all fuzzy truth values, that is,

$$L = \{a \mid a \text{ is a fuzzy set on } [0, 1]\}$$

For fuzzy truth values, three logic operations AND($\wedge$), OR($\vee$) and NOT($\neg$) are defined as follows by using the extension principle. Let a and b be fuzzy truth values and μ_a and μ_b be the membership functions of a and b, respectively. Then,

$$a \wedge b = \int \frac{\mu_a(x) \wedge \mu_a(y)}{(x \wedge y)} \tag{10}$$

$$a \vee b = \int \frac{\mu_a(x) \wedge \mu_b(y)}{(x \vee y)} \tag{11}$$

$$\neg a = \int \mu_a(x)/(1 - x) \tag{12}$$

where $x, y \in V$.

Two-valued truth values, numerical truth values and interval truth values are special cases of fuzzy truth values and the definitions of logic operations (1)–(3), (4)–(6) and (7)–(9) are special cases of the above definition (10)–(12), respectively.

Unfortunately, the set of all fuzzy truth values does not satisfy 3. Absorption laws and 5. Distributive laws, as follows.

Lemma 4: Fuzzy truth values do not satisfy the absorption laws.
Proof 4: For example, in fig. 3, $a \vee (a \wedge b)$ does not equal a. This is a counter example of the absorption laws.

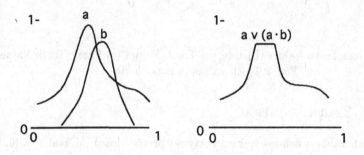

Fig. 3 A counter example of the absorption laws

Lemma 5: Fuzzy truth values do not satisfy the distributive laws.
Proof 5: For example, in fig 4, $a \vee (b \wedge c)$ does not equal $(a \vee b) \wedge (a \vee c)$. This is a counter example of the distributive laws.

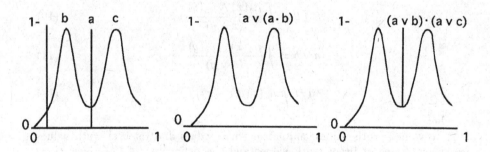

Fig. 4 A counter example of the distributive laws

The laws 1-4. are axioms of lattice. That is, fuzzy truth values do not satisfy even lattice, which is known as one of the simplest algebraic system.

Definition 4: A fuzzy set a is normal if and only if,

$$\bigvee_{x \in [0,1]} \mu_a(x) = 1.$$

Definition 5: A fuzzy set a is convex if and only if $\mu(y) \geq min(\mu(x), \mu(z))$ for all $x, y, z \in V$ such that $x \geq y \geq z$.

If we restrict fuzzy truth values such as normal and convex, that is,

$$L' = \{a|\ a \text{ is nomal and convex fuzzy set on } [0, 1]\}$$

then we can show the following theorem.

Theorem 3: The algebraic system $\langle L', \wedge, \vee, \neg \rangle$ is a de Morgan algebra (Proof Omitted).

5 Conclusion

In this paper, we clarified the fundamental properties of algebraic systems composed of truth values and three logic operations AND, OR and NOT which are used in fuzzy logic. We showed that numerical truth values satisfy the axioms of Kleene algebra, and interval truth values satisfy the axioms of De Morgan algebra, and normal and convex fuzzy truth values satisfy the axioms of de Morgan algebra[1] also, where de Morgan algebra is a weaker algebra than Kleene algebra and Kleene algebra is a weaker algebra than Boolean algebra.

References

1. Mukaidono, M., Kikuchi, H.: Fundamental properties of fuzzy interval logic. in: Advancement of Fuzzy Theory and Systems in China and Japan, International Academic Publishers (1990).
2. Zadeh, L.A.: The concept of linguistic variable and its application to approximate reasoning Part 1,2,3. Information Science Volume 8, Number 9 (1975,1976)

A Theory of Mass Assignments for Artificial Intelligence

J. F. Baldwin

Department of Computer Science, University of Bristol, England

Abstract. We present a theory of mass assignments for evidential reasoning under uncertainty which allows for fuzzy[15, 16, 17], probabilistic and incomplete probabilistic specifications[1, 2, 3, 4, 5, 6, 7, 8, 9, 10, 11]. The theory is applicable to fuzzy control, expert systems, decision support systems, knowledge engineering and represents a general theory of uncertainty in AI.

1 Review of Mass Assignment Theory

1.1 Definition of Mass Assignments

A mass assignment over a finite frame of discernment F is a function m, $m : P(F) \rightarrow [0,1]$ where $P(F)$ is the power set of F, such that,

$$\sum_{A \in P(F)} m(A) = 1; \ \forall A, \ m(A) \geq 0$$

If $m(\emptyset)$ then the mass assignment is said to be complete. Otherwise it is incomplete $m(A)$ for any $A \subseteq F$ represents a probability mass allocated to exactly the subset A of F.

Every subset, A, of F for which $m(A) > 0$ is called a focal element. If M is the set of focal elements of $P(F)$ for m then the mass assignment can be represented by $\{L_i : m_i\}$ where $L_i \in M$ and $m(L_i) = m_i$. This can also be written as $m = \{L_i : m_i\}$. We can also denote this mass assignment by (m, F) which we will often write simply as m. Total ignorance is then represented by the mass assignment $F : 1$. In this case the only focal element is F so that $m(F) = 1$ and $m(A) = 0$ for all subsets, A, of F other than F.

If $m = \{L_i : m_i\}$ then,

$$m' = \begin{cases} \{L_i : m_i | \ L_i \neq L_k, L_1', L_2'\} \cup \{L_1' : m(L_1') + x\} \cup \{L_2' : m(L_2') + y\} \\ \qquad \text{if } x + y = m_k \\ \{L_i : m_i | \ L_i \neq L_k, L_1', L_2'\} \cup \{L_1' : m(L_1') + x\} \cup \{L_2' : m(L_2') + y\} \\ \qquad \cup \{L_k : m_k - x - y\} \text{ if } m(L'1) + n(L'2) < m_k \end{cases}$$

is a <u>restriction</u> of m and is denoted by $m' \leq m$. Mass assignments $s1$ and $s2$ are said to be *orthogonal* if one cannot be obtained from the other by restriction.

The complement of a mass assignment $m = \{L_i : m_i\}$, denoted by $\overline{m}$, is $m = \{L_i' : m_i\}$ where the set of focal elements $F' = \{L_i'\}$ where L_i' is the complement of L_i with respect to F. If for m a mass is associated with F then this mass will be associated with $\emptyset$ for $\overline{m}$ and the complement is an incomplete mass assignment. If m is an incomplete mass assignment then the mass associated with $\emptyset$ is associated with F for $\overline{m}$.

1.2 General assignment method

Consider the two mass assignments (m_1, M_1) and (m_2, M_2) where $M_1 = \{L_{1k}\}$ for $k = 1, 2, ..., n_1$ and $M_2 = \{L_{2k}\}$ for $k = 1, 2, ..., n_2$ where each L_{ij} are subsets of F for which $m_1(L_{1k}) \neq \emptyset; k = 1, ..., n_1, m_2(L_{2k}) <> 0; k = 1, ..., n_2$. Let $*$ be a binary set theoretic operation, intersection or union. Let (m, M) be the result of combining (m_1, M_1) and (m_2, M_2) with respect to $*$ and we will denote this by,

$$(m(.|a), M) = (m_1, M_1) + *(m_2, M_2),$$

where m represents a unique assignment or family of mass assignments. If a unique assignment results then $m = m(.|\{\})$. The combination operation $*$ is defined by the general assignment algorithm now stated. Let $M = \{Lm_{ij}\}$ for which $Lm_{ij} = L1_i * L2_j$ for all i, j such that $L1_i * L2_j <> O$ and,

$$m(Y) = \sum_{i,j : L1_i * L2_j = Y} m'(L1_i * L2_j); \quad \text{for any } Y \subseteq M,$$

where $m'(L1_i * L2_j)$, for $i = 1, ..., n_1$ and $j = 1, ..., n_2$ satisfies,

$$\sum_j m'(L1_i * L2_j) = m1(L1_i) \text{ for } i = 1, ..., n_1,$$

$$\sum_i m'(L1_i * L2_j) = m2(L2_j) \text{ for } j = 1, ..., n_2,$$

$m'(L1_i * L2_j) = 0$ if $L1_i * L2_j = \emptyset$, for $i = 1, ..., n_1$ and $j = 1, ..., n_2$. Consider a matrix of cells C_{ij} where $i = 1, ..., n_1$ and $j = 1, ..., n_2$. Let cell C_{ij} contain label $L1_i * L2_j$ with associated mass $m'ij = m'(L1_i * L2_j)$. Then the i'th row masses of the cell tableau must add up to $m1(L1_i)$ for all i and the j'th column masses must add up to $m2(L2_j)$ for all j. Cells with null set entries are allocated the mass 0. This will not in general give a unique solution. For the non unique case, any allocation can be modified by alternatively adding and subtracting a quantity at vertices around a loop made up of alternative horizontal and vertical jumps. The quantity must be such that no cell mass entries go negative. This is simply the assignment algorithm from the field of operations research. If, in the case for intersection, the constraints cannot be satisfied, then the mass assignments are incompatible. An incomplete mass assignment for the combination can then be obtained by allowing masses to be associated with cells with null entries. If a unique solution is not obtained, the family of solutions can be parametrized. We will use the split algorithm defined below for obtaining the family of mass assignments rather than the loop method since the latter is difficult to handle computationally except for the more simple cases.

1.3 FMA's and basic operations

Let the frame of discernment be F and $\{s_1, \ldots, s_n\}$ be independent mass assignments defined over F. The linear form,

$$m = \sum_{i=1}^{n} a_i s_i; \quad \sum_{i=1}^{n} a_i = 1,$$

represents a family of mass assignments formed from the independent mass assignments $\{s_i\}$. We will call this a FMA.

Let M represent all FMA's over F. $\langle M, \vee, \wedge \rangle$ is an algebra with indempotence, commutativity, associativity, absorption, distributivity and complementation properties given above. Full complementation properties are not satisfied. The algebra is a pseudo Boolean Algebra. We can also view the structure in lattice terms. $\langle M, \leq \rangle$ is a poset and further is a lattice since the join and meet are defined everywhere. $\emptyset : 1$ and $F : 1$ are the universal bounds of the lattice M. The lattice is distributive but not completely complemented. $\langle M, \leq \rangle$ is a pseudo complemented distributed lattice.

1.4 Iterative Assignment Algorithm

The iterative assignment method for updating an *a priori* mass assignment with a sequence of evidences, each expressed as a mass assignment, is discussed by Baldwin [1]. We summarize the method. The update is a generalization of Bayesian updating to the case when the evidence is uncertain. It is in special cases equivalent to using Jeffrey's rule. It can also be viewed as a "filling in" process for the incompleteness expressed by the mass assignment. Suppose an *a priori* mass assignment ma is given over the focal set **A** whose elements are subsets of the power set $\mathcal{P}(F)$.

Suppose we also have a specific evidence E where E is (m, M) where M is the set of focal elements of $\mathcal{P}(F)$ for E and m is the mass assignment for these focal elements. We wish to update the *apriori* assignment m_a with E to give the updated mass assignment m. The iterative assignment method updates ma with E to give m'. The *apriori* is replaced with m' and the update process repeated to give a new m'. This is repeated until the process converges.

The one step algorithm is as follows:
$m = (t, T)$ where $t = \{t_1, \ldots, t_m\}, T = \{T_1, \ldots, T_m\}$, T_i is a subset of $\mathcal{P}(F)$.
i.e. $m = T_1 : t_1, \ldots, T_m : t_m$ and $E = (t^E, T^E)$ where $t^E = \{t_1^E, \ldots, t_s^E\}, T^E = \{T_1^E, \ldots, T_s^E\}$, T_i^E is a subset of $\mathcal{P}(F)$.
$E = T_1^E : t_1^E, \ldots, T_s^E : t_s^E$
$m' = (t', T')$ where $t' = \{t_1', \ldots, t_r'\}$ and $T' = \{T_1', \ldots, T_r'\}$
$T' = Set\{Bag\{T_i \cap T_j^E \mid T_i \cap T_j^E \neq \emptyset\}\}$

$$t_k' = \sum_{i,j : T_i \cap T_j^E = T_k'} K_j t_i t^E j \text{ for } k = 1, \ldots, r \text{ where}$$

$$K_j = \frac{1}{1 - \sum_{q:T_q \cap T_j^E = \emptyset} t_q} \quad \text{for } j = 1, ..., s$$

It should be noted that the label set, T', can change from stage to stage of the complete iteration process.

2 Fuzzy Sets and Mass Assignments

The theory of FMA's,[10] also includes fuzzy set theory as given in[15, 16, 17]. Let f be a normalized fuzzy set, $f \subseteq X$, such that,

$$f = \sum_{x_i \in X} x_i / \chi_f(x_i); \quad \chi_f(x_1) = 1, \quad \chi_f(x_K) \leq \chi_f(x_j) \ for \ k > j$$

where $\chi_f(x)$ is the membership level $x \in X$ of x in f.

The mass assignment associated with f is

$x_1 : 1 - \chi_f(x_2), \ \{x_1, ..., x_i\} : \chi_f(x_i) - \chi_f(x_{i+1})$ for $i = 2, ...$; with $\chi_f(x_k) = 0$ for $x_k \notin X$

This defines a family of probability distributions over F for the instantiation of variable X, given the statement X is f.

This definition is consistent with the voting model with the constant threshold model[9]. In this voting model $\chi_f(x)$ is the % of a representative population P who accept x as satisfying f with the constraint imposed on the voting behavior of P that any member of P who accepts $y \in X$ with membership value $\chi_f(y) \in f$, as satisfying f will also accept any $z \in X$ such that $\chi_f(z) \geq \chi_f(y)$.

Let f be an unnormalized fuzzy set, $f \subseteq X$, such that $max_i\{\chi_f(x_i)\} = \beta < 1$ where

$$f = \sum_{x_i \in X} x_i / \chi_f(x_i) \ ; \quad \chi_f(x_1) = \beta, \chi_f(x_k) \leq \chi_f(x_j) \text{ for } k > j$$

then the mass assignment associated with f is,

$x_1 : \beta - \chi_f(x_2), \{x_1, ..., x_i\} : \chi_f(x_i) - \chi_f(x_{i+1}), \vee$
$1 - \beta$ for $i = 2, ...$; with $\chi_f(x_k) = 0$.

3 Mass Assignment for Implication Statements

We will denote a random variable X with probability distribution belonging to a family of probability distributions defined by the mass assignment m as "X with m".

The rules IF X with m_{iX} THEN Y with m_{iY} for $i = 1, ..., n$ where $\{m_{iX}\}$ and $\{m_{iX}\}$ a re sets of mass assignments over F_I and F_O respectively, provide the general mass assignment,

$$m_{X \times Y} = \bigwedge_i (\overline{m_{iX}} \vee m_{iY}) = \bigwedge_i \overline{(m_{iX} \wedge \overline{m_{iY}})}$$

where $m_1 \wedge m_2$ represents the meet of m_1 and m_2, m_1/m_2 represents the join of m_1 and m_2, so that, for a given specific input m_X, we obtain the output

$$m_Y = Proj_Y [M_{X \times Y} \uparrow m_X]$$

where $m_1 \uparrow m_2$ is the mass assignment resulting from updating m_1 with m_2 using the iterative assignment algorithm.

A special case is that for the fuzzy rules IF $x \in f_i$ THEN $y \in g_i$ where $f_i \subseteq_f F_I$ and $g_i \subseteq_f F_O$ for $i = 1, \ldots, n$ which are replaced then by IF X with m_{ix} THEN Y with m_{iY} for $i = 1, \ldots, n$ where m_{iX} is the mass assignment associated with the fuzzy set g_i. Interpretations of the mass assignment rules will be given in the following examples.

4 Examples

4.1 Example 1

Consider the fuzzy set approximate reasoning example:

IF $x = f1 = a/1 + b/0.2$ THEN $y = g1 = \alpha/1 + \beta/0.3$
IF $x = f2 = a/0.1 + b/1 + c/0.2$ THEN $y = g2 = \alpha/0.1 + \beta/1 + g/0.3$
IF $x = f3 = b/0.2 + c/1$ THEN $y = g3 = b/0.1 + g/1$

These rules can be written in terms of mass assignments as follows:

If X with m_{f_1} THEN Y with m_{g_1}
if X with m_{f_2} THEN Y with m_{g_2}
if X with m_{f_3} THEN Y with m_{g_3}

where

$$m_{f_1} = a : 0.8, \{a, b\} : 0.2$$

$$m_{g_1} = \alpha : 0.7, \{\alpha, \beta\} : 0.3$$

$$m_{f_2} = b : 0.8, \{b, c\} : 0.1, \{a, b, c\} : 0.1$$

$$m_{g_2} = \beta : 0.7, \{\beta, \gamma\} : 0.2, \{\alpha, \beta, \gamma\} : 0.1$$

$$m_{f_3} = c : 0.8, \{b, c\} : 0.2$$

$$m_{g_3} = \gamma : 0.9, \{\beta, \gamma\} : 0.1$$

so that, the mass assignment for the rules is given by m, where,

$$m = m_1 \wedge m_2 \wedge m_3$$

where

$$m_i = (\overline{m_{f_i}} \vee m_{g_i}) = \overline{(m_{f_i} \wedge \overline{m_{g_i}})}.$$

Hence $\overline{m_{g_1}} = \{\beta, \gamma\} : 0.7, \gamma : 0.3,$
$\overline{m_{g_2}} = \{\alpha, \gamma\} : 0.7, \alpha : 0.2, \emptyset : 0.1,$
$\overline{m_{g_3}} = \{\alpha, \beta\} : 0.9, \alpha : 0.1.$

Therefore $m_{f_1} \wedge \overline{m_{g_1}} = a\gamma : 0.3, \{a\beta, a\gamma\} : 0.5, \{a\beta, a\gamma, b\beta, b\gamma\} : 0.2$ since,

	0.7 $\{\beta, \gamma\}$	0.3 γ
0.8 a	$\{a\beta, a\gamma\}$ 0.5	$a\gamma$ 0.3
0.2 $\{a, b\}$	$\{a\beta, a\gamma, b\beta, b\gamma\}$ max 0.2	$a\gamma, b\gamma$ 0

$m_{f_2} \wedge \overline{m_{g_2}} = b\,\emptyset : 0.1, b\alpha : 0.2, \{b\alpha, b\gamma\} : 0.5, \{b\alpha, b\gamma, c\alpha, c\gamma\} : 0.1, \{\_\alpha, \_\gamma\} : 0.1$ since,

	0.7 $\{\alpha, \gamma\}$	0.2 α	0.1 $\emptyset$
0.8 b	$\{b\alpha, b\gamma\}$ 0.5	$b\alpha$ 0.2	$\emptyset$ 0.1
0.1 $\{b, c\}$	$\{b\alpha, b\gamma, c\alpha, c\gamma\}$ max 0.1	$\{b\alpha, c\alpha\}$ 0	$\emptyset$ 0
0.1 $\{\_\}$	$\{\_\alpha, \_\gamma\}$ 0.1 max	$\{\_\alpha\}$ 0	$\emptyset$ 0

$m_{f_3} \wedge \overline{m_{g_3}} = c\alpha, 0.1, \{c\alpha, c\beta\} : 0.7, \{b\alpha, b\beta, c\alpha, c\beta\} : 0.2$ since,

	0.9 $\{\alpha, \beta\}$	0.1 α
0.8 c	$\{c\alpha, c\beta\}$ 0.7	$c\alpha$ 0.1
0.2 $\{b, c\}$	$\{b\alpha, b\beta, c\alpha, c\beta\}$ max 0.2	$b\alpha, c\alpha$ 0

so that $m_1 = \overline{m_{f_1} \wedge \overline{m_{g_1}}} = \{a\alpha, a\beta, b_-, c_-\} : 0.5, \{a\alpha, b\alpha, c_-\} : 0.2$
$m_2 = \overline{m_{f_2} \wedge \overline{m_{g_2}}} = \{\_, \_\} : 0.1, \{a_-, b\beta, b\gamma, c_-\} : 0.2, \{a_-, b\beta, c_-\} : 0.5, \{a_-, b\beta, c\beta\}; 0.1, \{\_\beta\} \, 0.1$
$m_3 = \overline{m_{f_3} \wedge \overline{m_{g_3}}} = \{a_-, b_-, c\beta, c\gamma\} : 0.1, \{a_-, b_-, c\gamma\} : 0.7 \{a_-, b\gamma, c\gamma\} : 0.2.$

28

Therefore $m_1 \wedge m_3 = \{a\alpha, c\gamma\} : 0.2, \{a\alpha, b_-, c\gamma\} : 0.5, \{a\alpha, a\beta, b_-, c\gamma\} : 0.2, \{a\alpha, a\beta, b_-, c\beta, c\gamma\} :$
0.1 since

	0.1 {a_, b_, cβ, cγ}	0.7 {a_, b_, cγ}	0.2 {a_, bγ, cγ}
0.3 {aα, aβ, b_, c_}	{aα, aβ, b_, cβ, cγ} max1 0.1	{aα, aβ, b_, cγ} max2 0.2	{aα, aβ, bγ, cγ} 0
0.5 {aα, b_, c_}	{aα, b_, cβ, cγ} 0	{aα, b_, cγ} max3 0.5	{aα, bγ, cγ} 0
0.2 {aα, bα, c_}	{aα, bα, cβ, cγ} 0	{aα, bα, cγ} 0	{aα, cγ} 0.2

$(m_1 \wedge m_3) \wedge m_2 = m_1 \wedge m_2 \wedge m_3 = a\alpha, 0.1, b\beta : 0.1, \{a\alpha, c\gamma\} : 0.1, \{a\alpha, b\beta, c\gamma\} :$
$0.4, \{a\alpha, b\beta, b\gamma\} : 0.2, \{a\alpha, a\beta, b_-, c\beta, c\gamma\}; 0.1$ since

	{aα, cγ}	{aα, b_, cγ}	{aα, aβ, b_, cγ}	{aα, aβ, b_, cβ, cγ}
0.1 {_ _}	{aα, cγ} 0	{aα, b_, cγ} 0	{aα, aβ, b_, cγ} 0	{aα, aβ, b_, cβ, cγ} max1 0.1
0.2 {a_, bβ, bγ, c_}	{aα, cγ} 0	{aα, bβ, bγ, cγ} 0	{aα, aβ, bβ, bγ, cγ} max2 0.2	{aα, aβ, bβ, bγ cβ, cγ} 0
0.5 {a_, bβ, c_}	{aα, cγ} 0.1	{aα, bβ, cγ} max3 0.4	{aα, aβ, bβ, cγ} 0	{aα, aβ, bβ, cβ, cγ} 0
0.1 {a_, bβ, cβ}	aα 0.1	{aα, bβ} 0	{aα, aβ, bβ} 0	{aα, aβ, bβ, cβ} 0
0.1 {_β}	∅ 0	bβ 0.1	{aβ, bβ} 0	{aβ, bβ, cβ} 0

If a specific input X with $m_X = a : 0.4, b : 0.1, c : 0.5$ is given then the associated inferred specific output is

Y with $Proj_Y\{m \uparrow m_x\} = a\alpha : 0.2667, b\beta : 0.0625, c\gamma : 0.4375, \{a\alpha, a\beta\} : 0.1333, \{b\beta, b\gamma\} : 0.025, \{b_-\} : 0.0125, \{c\beta, c\gamma\} : 0.0625$

This converges in one step using the iterative assignment algorithm. We can obtain a further "fill in" using,

Y with $Proj_Y\{e_{X \times Y} \uparrow (m \uparrow m_x)\} = \alpha : 0.4714, \beta : 0.0143, \gamma : 0.5143$

The following "fill in" schemes will only give approximate solutions,

$Proj_Y\{e_{X \times Y} \uparrow m, m_x)\} = \alpha : 0.4, \beta : 0.1, \gamma : 0.5$

since, this requires that both m and m_x is satisfied.

$e_Y \uparrow Proj_Y\{m \uparrow m_x\} = \alpha : 0.3746, \beta : 0.1093, \gamma : 0.5162$

since equally likely distribution taken over only the Y values corresponds in this case to an unequally likely distribution over the cross product space of X and Y.

The solution given here for the mass assignment approach is slightly different from that given for the fuzzy logic approach. This difference arises because the mass assignment method found a solution which is compatible with all the rules while the fuzzy logic approach results in a non-normalised fuzzy set corresponding to some inconsistency. This solution could be obtained from the last tableau if we allowed a mass to be associated with the cell with label $\emptyset$. The resulting tableau would then be that given below and the resulting mass assignment $a\alpha, 0.1, \{a\alpha, b\beta, c\gamma\} : 0.5, \{a\alpha, a\beta, b\beta, b\gamma, c\gamma\} : 0.2, \{a\alpha, a\beta, b_, c\beta, c\gamma\} : 0.1, \emptyset : 0.1$ would correspond to the fuzzy set $a\alpha/0.9 + b\beta/0.8 + c\gamma/0.8 + a\beta/0.3 + b\gamma/0.3 + ba/0.1 + c\beta/0.1$ which corresponds to that given by the fuzzy logic approach.

	{aα, cγ}	{aα, b_, cγ}	{aα, aβ, b_, cγ}	{aα, aβ, b_, cβ, cγ}
0.1 {_ _}	{aα, cγ} 0	{aα, b_, cγ} 0	{aα, aβ, b_, cγ} 0	{aα, aβ, b_, cβ, cγ} max1 0.1
0.2 {a_, bβ, bγ, c_}	{aα, cγ} 0	{aα, bβ, bγ, cγ} 0	{aα, aβ, bβ, bγ, cγ} max2 0.2	{aα, aβ, bβ, bγ cβ, cγ} 0
0.5 {a_, bβ, c_}	{aα, cγ} 0	{aα, bβ, cγ} max3 0.5	{aα, aβ, bβ, cγ} 0	{aα, aβ, bβ, cβ, cγ} 0
0.1 {a_, bβ, cβ}	aα max 0.1	{aα, bβ} 0	{aα, aβ, bβ} 0	{aα, aβ, bβ, cβ} 0
0.1 {_β}	∅ 0.1	bβ 0	{aβ, bβ} 0	{aβ, bβ, cβ} 0

The mass assignment compatible with the rules is to be preferred to that corresponding to the fuzzy logic approach. From a practical point of view, considering that the fuzzy logic solution is only slightly different, we would prefer the fuzzy logic solution since the computation is much less. For those examples in which the fuzzy logic approach gives a the solution as a normalized fuzzy set, the mass assignment method will give an equivalent solution. For the case in which the fuzzy logic solution is non-normalised, the two methods will be exactly equivalent only if the mass assignment method cannot give a compatible solution.

4.2 Example 2

Our second example is that given by Pearl[13] to illustrate his theory of causal nets. We apply the general mass assignment theory to this example. This extends the solution for cases involving incompleteness. We introduce the concept of independent mass assignments which has been discussed by Baldwin. For this paper we simple note that if the meet of independent mass assignments is obtained by multiplying the masses for each cell of the split algorithm. Row and column constraints must still be satisfied and this is done by re-normalisation if

Ø sets occur. This is illustrated below. It does not correspond to Dempster's-rule.

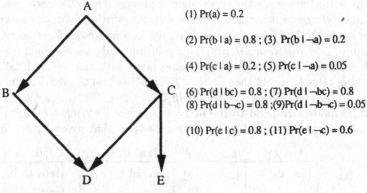

(1) Pr(a) = 0.2

(2) Pr(b | a) = 0.8 ; (3) Pr(b | ¬a) = 0.2

(4) Pr(c | a) = 0.2 ; (5) Pr(c | ¬a) = 0.05

(6) Pr(d | bc) = 0.8 ; (7) Pr(d | ¬bc) = 0.8
(8) Pr(d | b¬c) = 0.8 ;(9)Pr(d | ¬b¬c) = 0.05

(10) Pr(c | c) = 0.8 ; (11) Pr(e | ¬c) = 0.6

PHASE 1 Forward Computation of Apriori's
Computation of apriori mass assignment over ABC
Mass Assignments corresponding to (1) to (5) above
m1 = a 0.2, ¬a : 0.8
m2 = {ab, ¬a_} : 0.8, {a¬b, ¬a_} : 0.2
m3 = {a_, ¬ab} : 0.2, {a_, ¬a¬b} : 0.8
m4 = {ac, ¬a_} : 0.2, {a¬c, ¬a_} : 0.8
m5 = {a_, ¬ac} : 0.05, {a_, ¬a¬c} : 0.95

The conditional independence constraint, $Pr(BC|A) = Pr(B|A)Pr(C|A)$, of causal nets is equivalent to the constraints that $m2$ and $m4$ are independent, and $m3$ and $m5$ are independent. Therefore, since $m2$ and $m4$ are independent $m2 \wedge m4 = \{abc, \neg a_-, _-\} : 0.16, \{ab\neg c, \neg a_-, _-\} : 0.64, \{a\neg bc, \neg a_-, _-\} : 0.04, \{a\neg b\neg c, \neg a_-, _-\} : 0.16$ and $m1 \wedge (m2 \wedge m4) = abc : 0.032, ab\neg c : 0.128, a\neg bc : 0.008, a\neg b\neg c : 0.032, \{\neg a_-, _-\} : 0.8$ using independence assumption for m_1 and $(m_2 \wedge m_4)$, since

	0.2 a	0.8 ¬a
0.16 {abc, ¬a_ _}	abc 0.032	{¬a_ _} 0.128
0.64 {ab¬c, ¬a_ _}	ab¬c 0.128	{¬a_ _} 0.512
0.04 {a¬bc, ¬a_ _}	a¬bc 0.008	{¬a_ _} 0.032
0.16 {a¬b¬c, ¬a_ _}	a¬b¬c 0.032	{¬a_ _} 0.128

Similarly m_3 and m_5 are independent so that $m_3 \wedge m_5 = \{a_-, _-, \neg abc\} : 0.01, \{a_-, _-, \neg ab\neg c\} : 0.19, \{a_-, _-, \neg a\neg bc\} : 0.04, \{a_-, _-, \neg a\neg b\neg c\} : 0.76$ Also m_1 and $(m_3 \wedge m_5)$ are independent, so that $m_1 \wedge m_3 \wedge m_5 = m_1 \wedge (m_3 \wedge m_5) = \neg abc : 0.008, \neg ab\neg c : 0.152, \neg a\neg bc : 0.032, \neg a\neg b\neg c : 0.608, \{a_-\} : 0.2$. Therefore $m_1 \wedge m_2 \wedge m_3 \wedge m_4 \wedge m_5 = (m_1 \wedge m_2 \wedge m_4) \wedge (m_1 \wedge m_3 \wedge m_5) = abc : 0.032, ab\neg c : 0.128, a\neg bc : 0.008, a\neg b\neg c : 0.032, \neg abc : 0.008, \neg ab\neg c : 0.152, \neg a\neg bc : 0.032,$

$\neg a \neg b \neg c : 0.608 = a\ priori$ mass assignment over $ABC = m$ say, since

	¬abc	¬ab¬c	¬a¬bc	¬a¬b¬c	{a__}
0.032 abc	∅ 0	∅ 0	∅ 0	∅ 0	abc 0.032
0.128 ab¬c	∅ 0	∅ 0	∅ 0	∅ 0	ab¬c 0.128
0.008 a¬bc	∅ 0	∅ 0	∅ 0	∅ 0	a¬bc 0.008
0.032 a¬b¬c	∅ 0	∅ 0	∅ 0	∅ 0	a¬b¬c 0.032
0.8 {¬a__}	¬abc 0.008	¬ab¬c 0.152	¬a¬bc 0.032	¬a¬b¬c 0.608	∅ 0

4.3 Computation of a priori mass assignment over BCD

Projecting the mass assignment over ABC computed above onto BC gives the *a priori* mass assignment over BC as $n1 = Proj_{BC}m = bc : 0.04, \neg bc : 0.04, b \neg c : 0.28, \neg b \neg c : 0.64$

Mass assignments corresponding to (6), (7), (8), (9) are given by

$$n2 = \{bcd, b \neg c_{\neg}, \neg b_{\_}\} : 0.8, \{bc \neg d, b \neg c_{\neg}, \neg b_{\_}\} : 0.2$$

$$n3 = \{\neg bcd, \neg b \neg c_{\neg}, b_{\_}\} : 0.8, \{\neg bc \neg d, \neg b \neg c_{\neg}, b_{\_}\} : 0.2$$

$$n4 = \{b \neg cd, bc_{\neg}, \neg b_{\_}\} : 0.8, \{b \neg c \neg d, bc_{\neg}, \neg b_{\_}\} : 0.2$$

$$n5 = \{\neg b \neg cd, \neg bc_{\neg}, b_{\_}\} : 0.05, \{\neg b \neg c \neg d, \neg bc_{\neg}, b_{\_}\} : 0.95$$

The *a priori* conditional constraint implies that $n1$ and $n2$ are independent, $n1$ and $n3$ are independent, $n1$ and $n4$ are independent and $n1$ and $n5$ are independent, so that

$$n1 \wedge n2 = bcd : 0.032, bc \neg d : 0.008, \neg bc_{\_} : 0.04, b \neg c_{\_} : 0.28, \neg b \neg c_{\_} : 0.64$$

$$n1 \wedge n3 = \neg bcd : 0.032, \neg bc \neg d : 0.008, bc_{\_} : 0.04, b \neg c_{\_} : 0.28, \neg b \neg c_{\_} : 0.64$$

$$n1 \wedge n4 = b \neg cd : 0.224, b \neg c \neg d : 0.056, bc_{\_} : 0.04, \neg bc_{\_} : 0.04, \neg b \neg c_{\_} : 0.64$$

$$n1 \wedge n5 = \neg b \neg cd : 0.032, \neg b \neg c \neg d : 0.608, bc_{\_} : 0.04, \neg bc_{\_} : 0.04, b \neg c_{\_} : 0.28$$

Therefore the split algorithm for conjunction gives $n1 \wedge n2 \wedge n3 = (n1 \wedge n2) \wedge (n1 \wedge n3) = bcd : 0.032, bc \neg d : 0.008, \neg bcd : 0.032$, and $\neg bc \neg d : 0.008, b \neg c_{\_} : 0.28, \neg b \neg c_{\_} 0.64$ and $n1 \wedge n4 \wedge n5 = (n1 \wedge n4) Y (n1 \wedge n5) = b \neg cd : 0.224, b \neg c \neg d : 0.056, \neg b \neg cd : 0.032, \neg b \neg c \neg d : 0.608, \neg bc_{\_} : 0.04, bc_{\_} : 0.04$

so that again using the split algorithm for conjunction

$n1 \wedge n2 \wedge n3 \wedge n4 \wedge n5 = (n1 \wedge n2 \wedge n3) \wedge (n1 \wedge n4 \wedge n5) = bcd : 0.032, bc \neg d : 0.008, b \neg cd : 0.224, b \neg c \neg d : 0.056, \neg bcd : 0.032, \neg bc \neg d : 0.008, \neg b \neg cd : 0.0032, \neg b \neg c \neg d : 0.608 = n$ say

which is the a priori mass assignment over BCD.

4.4 Computation of a priori mass assignment over CE

Projecting the mass assignment over BCD on to C gives the a priori mass assignment over C. i.e $r1 = Proj_{C}n = c : 0.08, \neg c : 0.92$ Mass assignments corresponding to (10) and (11) are given by $r2 = \{ce, \neg c_-\} : 0.8, \{c \neg e, \neg c_-\} : 0.2$; $r3 = \{c_-, \neg ce\} : 0.6, \{c_-, \neg c \neg e\} : 0.4$. *Apriori*/conditional independence condition constrains $r1$ and $r2$ to be independent and also $r1$ and $r3$ to be independent. So that $r1 \wedge r2 = ce : 0.064, c \neg e : 0.016, \neg c_- : 0.92$ and $r1 \wedge r3 = \neg ce : 0.552, \neg c \neg e : 0.368, \neg c_- : 0.08$. The split algorithm for conjunction gives

$$r1 \wedge r2 \wedge r3 = (r1 \wedge r2) \wedge (r1 \wedge r3) = ce : 0.064, c \neg e : 0.016,$$

$$\neg ce : 0.552, \neg c \neg e : 0.368 = r \text{ say}$$

This is the *apriori* mass assignment over CE.

4.5 Computation of apriori mass assignment over $BCDE$

$Pr(DE|C) = Pr(D|C) \cdot Pr(E|C)$, ie $D|E$ and $C|E$ are independent implies that the mass assignments calculated above, namely n and r are independent. Therefore the apriori mass assignment over $BCDE$, nr say, is given by $nr = n \wedge r$ determined using independence constraint. Therefore,

$nr = bcde : 0.0256, bcd \neg e : 0.0064, bc \neg de : 0.0064, bc \neg d \neg e : 0.0016 b \neg cde : 0.1344, b \neg cd \neg e : 0.0896, b \neg c \neg de : 0.0336, b \neg c \neg d \neg e : 0.0224 \neg bcde : 0.0256, \neg bcd \neg e : 0.0064, \neg bc \neg de : 0.0064, \neg bc \neg d \neg e : 0.0016 \neg b \neg cde : 0.0192, \neg b \neg cd \neg e : 0.0128, \neg b \neg c \neg de : 0.3648, \neg b \neg c \neg d \neg e : 0.2432$ since

	0.064 ce	0.016 c¬e	0.552 ¬ce	0.368 ¬c¬e	
0.032 bcd	bcde 0.002048k1	bcd¬e 0.000512k2	Ø 0	Ø 0	
0.008 bc¬d	bc¬de 0.000512k1	bc¬d¬e 0.000128k2	Ø 0	Ø 0	NOTE Row and
0.224 b¬cd	Ø 0	Ø 0	b¬cde 0.123648k3	b¬cd¬e 0.082432k4	column constraints are satisfied
0.056 b¬c¬d	Ø 0	Ø 0	b¬c¬de 0.030912k3	b¬c¬d¬e 0.020608k4	by renormalising so that each
0.032 ¬bcd	¬bcde 0.002048k1	¬bcd¬e 0.000512k2	Ø 0	Ø 0	column constraint is satsified.
0.008 ¬bc¬d	¬bc¬de 0.000512k1	¬bc¬d¬e 0.000128k2	Ø 0	Ø 0	The row constraints are also satisfied
0.032 ¬b¬cd	Ø 0	Ø 0	¬b¬cde 0.017664k3	¬b¬cd¬e 0.011776k4	when a compatible solution with
0.608 ¬b¬c¬d	Ø 0	Ø 0	¬b¬c¬de 0.335616k3	¬b¬c¬d¬e 0.223744k4	both masses exist.

$k1 = 0.064 / 0.00512 \quad k2 = 0.016 / 0.00128 \quad k3 = 0.552 / 0.50784 \quad k4 = 0.368 / 0.33856$

4.6   PHASE 2 Backward Computation of Updated Mass Assignments

Suppose we have specific mass assignments for D and for E i.e.,

$$s1 = d : \alpha, \neg d : \beta, \{d, \neg d\} : 1 - \alpha - \beta; \quad s2 = e : g, \neg e : d, \{e, \neg e\} : 1 - \gamma - \delta$$

The following updating computations are performed to obtain the updated mass assignment for A.

$$nr' = nr \uparrow s1, s2; \quad m'_{BC} = Proj_{BC} nr'; m'_{ABC} = m \uparrow m'_{BC}; \quad m'_A = Proj_A m'_{ABC}$$

so that the updated mass assignment for A is

$$a : \xi, \neg a : \varsigma \{a, \neg a\} : 1 - \xi - \varsigma$$

Some results are For $s1 = \neg d : 1, s2 = e : 1$ then the update mass assignment is $a : 0.0934, \neg a : 0.9066$ For $s1 = d : 0.1, \neg d : 0.9, s2 = e : 0.9, \neg e : 0.1$ then the update mass assignment is $a : 0.1297, \neg a : 0.8703$.

5   Conclusions

A general theory for evidential reasoning under uncertainty has been presented. It handles both fuzzy and probabilistic uncertainties. The examples given shows its application to fuzzy control and causal nets. We could also have illustrated application to database theory in which attribute values can be fuzzy sets or probability distribution or mass assignments, non-monononic logic problems. Theorems of decomposition exist for simplifying computations. An obvious example is given in example 2 above where both for the forward and backward phases the problem is split into 2 parts, that applicable to ABC and to BCDE. The iterative assignment method extends the concept of Bayes' conditioning to updating mass assignments under uncertain evidence. In simple cases this is equivalent to the theory of conditioning under uncertainty given by Jeffreys. It is also related to maximum entropy methods which has been discussed by Baldwin in references given below.

References

1. Baldwin J.F.: A New Approach to Combining Evidences for Evidential Reasoning. ITRC Univ. of Bristol Report, (1989).
2. Baldwin J.F.: Computational Models of Uncertainty Reasoning in Expert Systems. Computers Math. Applic., Vol. 19, No 11, (1990), pp 105-119.
3. Baldwin J.F.: Combining Evidences for Evidential Reasoning", Int. J. of intelligent Systems, (1990).
4. Baldwin J.F.,: Towards a general theory of intelligent reasoning. 3rd International Conference IPMU, Paris, (1990).

5. Baldwin J.F.: Evidential Reasoning under Probabilistic and Fuzzy Uncertainties. in An introduction to Fuzzy Logic Applications in Intelligent Systems (Ed. R. R. Yager and L. A. Zadeh), Kluwer Academic Publishers, (1990).

6. Baldwin J, F.: Inference under uncertainty for Expert System Rules. ITRC 152 University of Bristol Report (1990).

7. Baldwin J.F.: Inference for Information Systems containing Probabilistic and Fuzzy Uncertainties. in Fuzzy Logic for the Management of Uncertainty, Ed. Lotfi L. Zadeh and Janusz Kacprzyk, Wiley & Sons, (1990).

8. Baldwin J.F.: The management of fuzzy and probabilistic uncertainties for knowledge based systems. ITRC Univ of Bristol Report No. 154, To appear in AI Encyclopedia, (1990).

9. Baldwin J. F. : Fuzzy and Probabilistic DataBases with Automatic Reasoning. ITRC 160, University of Bristol Report, (1991).

10. Baldwin J. F.: A calculus for Mass Assignments in Evidential Reasoning. ITRC 163, University of Bristol Report, in Advances in the Dempster-Shafer Theory of Evidence, (Ed Mario Fedrizza, Janusz Kacprzyk and Ronald Yager), Wiley & Sons, Inc. New York (1991).

11. Baldwin J.F.: Mass Assignments and fuzzy sets for fuzzy databases. ITRC 162, University of Bristol Report, in Advances in the Dempster-Shafer Theory of Evidence, (Ed Mario Fedrizza, Janusz Kacprzyk and Ronald Yager), Wiley & Sons, Inc. New York (1992).

12. Jeffrey R.: The Logic of Decision, McGraw-Hill, New York (1965).

13. Pearl J.: Probabilistic reasoning in Intelligent Systems, Morgan Kaufmann, (1988).

14. Shafer G.: A mathematical theory of evidence. Princeton University Press, (1976).

15. Zadeh, L: Fuzzy sets. Information and Control, 8, (1965), pp 338-353.

16. Zadeh, L.: Fuzzy logic and Approximate Reasoning. Synthese, 30, (1975), pp 407-428.

17. Zadeh, L.: Fuzzy Sets as a basis for a theory of Possibility. Fuzzy Sets and Systems 1, (1978), 3-28

Part II

Fuzzy Logic-based Reasoning and Control: Theoretical Aspects - Fuzzy Control

Fuzzy Dynamic Systems

W. Pedrycz

Department of Electrical and Computer Engineering, University of Manitoba,
Winnipeg, Manitoba, Canada R3T 2N2

Abstract. The aim of this paper is to study fuzzy dynamic systems.
The role of fuzzy sets in formation of a conceptual and computational
platform for symbolic and numerical information processing is identified.
We summarize essential properties of fuzzy partitions developed with the
aid of families of fuzzy sets. The problem of modelling with fuzzy sets is
addressed with respect to fuzzy partitions defined for system's variables.
Interesting trade-offs between essential features of fuzzy models (such
as precision and generality) induced by fuzzy partitions applied to the
model are also pointed out. In this context we highlight conceptual links
between fuzzy modelling and Qualitative Modelling. Main classes of fuzzy
models are introduced.

Keywords: fuzzy models, qualitative modelling, relational structures.

1 Introduction

Fuzzy sets emerged as formal models used to describe concepts with imprecisely
defined boundaries[9]. Their use in system modelling which gave rise to a class
of fuzzy dynamical models has far reaching methodological consequences. Fuzzy
sets are placed between methods of numerical and symbolic information processing and so are fuzzy models occupying an important niche distinct from well
established areas of purely qualitative and quantitative modelling. The aim of
the paper is to introduce and study some essential properties of fuzzy dynamic
models. The two evident advantages of fuzzy modelling concern a high flexibility of resulting models and their abilities to represent and cope with uncertain
information. Fuzzy sets will be investigated with respect to selected aspects of
knowledge representation. Those deal with syntax and semantics of objects cap-
tured by fuzzy sets and information granularity defined in this manner. The
term of a fuzzy partition developed there will be recognized as a key factor in
expressing a variable level of generality of constructed models. The paradigm of
fuzzy modelling discussed in Section 3 will show what could be accomplished in
this area of modeling. Classes of dynamic models will be carefully investigated
in Section 4.

2 Fuzzy sets: knowledge representation issues

2.1 Fuzzy sets: numerical and symbolic processing

Fuzzy sets are usually viewed as formal mathematical objects describing concepts within imprecisely defined boundaries. By admitting different grades of

membership of the elements of the universe over which a fuzzy set is defined, we can interpret it as a linguistic label used in representing knowledge about a domain of a given problem[10]. Fuzzy sets convey both symbolic and numerical information. When we look globally at them as formal objects, each label can be treated as a symbol. Looking closer at the elements of the object we can also distinguish more numerical details, since a membership function of a fuzzy set provides a detailed numerical information (numerical values of grades of membership). Thus, in addition to their symbolic nature, fuzzy sets are well defined in terms of their semantics. It is worth noting that the set theory (where sets are defined by two-valued characteristic functions) treats its objects in a different way. The elements of the same set are nondistinguishable (i.e., each two $x, y \in A$ are equivalent in the sense of the constraint defined by this set). Thus the equivalence relation plays a dominant role in the set theory. Fuzzy sets introduce a significant relaxation accepting the use of similarity relations in place of the Boolean equivalence. The numerical nature of fuzzy sets makes it possible to process them at a numerical level. Its symbolic character allows to carry out symbolic calculations.

2.2   Information granularity and fuzzy partition

Fuzzy sets treated as linguistic labels or information granules[11] structure elements of a given universe X (e.g., numerical values) into several general categories. These categories are put together and form a frame of cognition or a fuzzy partition of X. More formally a family of fuzzy sets;

$$\{A_1, A_2, \ldots, A_c\}$$

where $A_i : X \rightarrow [0,1]$) constitutes a frame A if the following three properties are satisfied:

– a "covers" the universe X, namely each element of the universe is assigned to at least one granula with a nonzero degree of membership:

$$\forall_x \exists_i A_i(x) > 0$$

– the coverage property means that any piece of information in X can be properly represented by A_i's.
– The elements of A are unimodal fuzzy sets. By stating that we identify several regions of X (one for each A_i) highly compatible with the labels (i.e., with higher grades of membership in A_i).

Below we list a series of essential features of the fuzzy partition:

Granularity Each frame A has a certain granularity which implies an amount of details represented and distinguished there. Considering a threshold level λ ($\lambda \in (0,1)$) each label A_i is converted into a set including elements satisfying the concept A_i to a degree not lower than λ (a λ-cut of A_i),

$\{x|A_i(x) \geq \lambda\}$. For sufficiently high values of λ the obtained λ-cuts are disjoint. The higher the cordinality of A, $card(A) = c$, the higher the specificity of the frame A. The brittleness of the linguistic terms A_i diminishes as this number (c) increases.

Information hiding By using this concept (originating from software engineering) we make some elements of X nondistinguishable and thus treat them as equivalent. In other words, we "hide" them and allow a collection of objects to be perceived and processed as a single element. Information hiding is completed on purpose so that all following computational processes will not be carried out below this conceptual level. We have already shown that this an inherent property of the set theory. Fuzzy sets allow to add on extra flexibility to this term by parameterizing it along allowable grades of membership. In other words, λ-cuts are sets completing information hiding at a specified level. Particularly when trapezoidal fuzzy numbers are used to construct the frame of congnition, their λ-cuts with $\lambda = 1$ imply that the information hiding is performed at its highest level.

Robustness Each fuzzy partition is characterized by a certain robustness which is partially determined by the level of information granularity of the partition. The idea states that fuzzy sets provide more robustness and can tolerate incorrect (noisy) information. Generally speaking this is due to a smooth rather than abrupt transition from complete membership to complete exclusion expressed by a membership function. Experiments reveal that the fuzzy partition is characterized by a higher robustness in comparison to that characterizing the induced Boolean partition (i.e., obtained from the fuzzy partition). General findings show that the robustness is strongly influenced by the specificity of the linguistic terms; usually lower specificity causes higher robustness. The above properties of the fuzzy partition will have a primordial impact on the fuzzy modeling yielding models of variable generality driven exclusively by the initial frame of cognition.

Before we will study fuzzy models one should conclude that fuzzy partition leads to a homogeneous form of information available to fuzzy models. To elaborate a bit on this issue let us note that an input information can be uniquely expressed in terms of linguistic labels. The converse is not true; knowing the representative of the input information at the level of the information granules one cannot reconstruct it in a unique manner. This phenomenon is due to the level of generality expressed by fuzzy sets. The scheme of transformation can be portrayed qualitatively as follows:

input information $\rightarrow$ fuzzy partition $\rightarrow x$

where the input information could be given in a numerical, interval, or fuzzy set format. The derived output x expresses this input by providing degrees of matching (activation) of elements in the fuzzy partition. Thus x becomes associated with the fuzzy partition and its semantics directly reflects the semantics of the linguistic labels. In a simple example where the input information is given precisely as a single numerical quantity x_0, the vector $x \in [0,1]^c$ can include possibility measures of x_0 taken with respect to $A_1, A_2, \ldots, A_c$, namely,

$$x = [A_1(x_0), A_2(x_0), \ldots, A_c(x_0)].$$

For interval-valued information $[x_0, x_{00}]$ the resulting vector x is expressed as;

$$x = \sup A_1(x) \quad \sup A_2(x) \quad \ldots \quad \sup A_c(x) \quad \text{for } x \in [x_0, x_{00}].$$

Thus the proposed transformation yields homogeneous output results despite a possible diversity of available inputs.

3 System modelling with fuzzy sets

3.1 The paradigm of fuzzy modelling

The basic idea of fuzzy models and fuzzy modelling is to model systems at a level of linguistic labels. The models of this class are not used to represent relationships between variables at a numerical, pointwise level. Their role is to look at the system by accepting a proper cognitive perspective. The fuzzy partition constructed for each variable can be adjusted separately to meet requirements of the modelling task. Some details can be then selectively hidden and won't increase unnecessary computational burden of model building. If more details are required one can formulate another more detailed fuzzy partition and re-design the fuzzy model. The fuzzy models developed in this way concur with the principle of incompatibility formulated by Zadeh[10]; a similar formulation can be found in [7]. The principle states that any model building calls for a rational trade-off between significance (relevance) and precision achievable within the model. It should be stressed that one should sacrifice (to a certain degree) precision to reach an acceptable level of generality. Overall the properties of the frame of cognition applied to modelling are in-built into models. For instance, increasing robustness of the frame A implies higher robustness of the fuzzy model. The identification will be therefore concentrated on expressing links between linguistic labels. In sequel we will be concerned with constructing formal relationships describing this dynamics.

3.2 Linguistic dynamic models

In this section we will develop a generic class of models. Since they operate at the level of linguistic labels they will be referred to as linguistic models. The dynamics of the system described linguistically is formulated by giving a relational description R between u_k, x_k and x_{k+1}.

$$(u_k, x_k, x_{k+1}) \in R \tag{1}$$

u_k is expressed in the frame of the input (control) variable and x_k as well as x_{k+1} describe a location of the state variable in successive time moments. While this expression is plausible from a conceputal point of view, further calculations require more algorithmic details of the description. We can rewrite (1) as,

$$x_{k+1} = (u_k \times x_k) \circ R, \tag{2}$$

where a Cartesian product of u_k and x_k (expressing a disjunctive form of available knowledge about a current state of the system) is propagated through R. The composition operator can be selected in several ways. The one being frequently utilized assumes a logic-based character of constraints and their propagation. This leads us towards an OR_AND combination carried out over R and the current characterization of the state of the system $u_k \times x_k$, say z_k,

$$x_{k+1} = OR\_AND(z_k, R). \tag{3}$$

Coordinatewise this formula reads as,

$$x_{k+1}(x_i) = OR \ [z_k(z_1) \ AND \ R(z_1, x_i), z_k(z_2) \ AND \ R(z_2, x_i) \ldots],$$

where z_i are pairs of u_j and x_k over which the Cartesian product $u_k \times x_k$ is defined. The above formula can be interpreted as an optimistic aggregation of the constraints $(z_k(z_j) \ AND \ R(z_j, x_i))$.

The logical operators (AND and OR) are usually modelled by triangular norms [4]. An important class of OR_AND composition is constituted by the maximum and t-norms. The general case involves s-norms (OR operator) and t-norms (AND operator).

The linguistic model described by (2) is shown in Fig. 1 The matrix of connections realizes the Cartesian product $u_k \times x_k$ (the product is usually implemented with the use of t-norms). The second block AND's its outputs and the corresponding entries of the fuzzy relation.

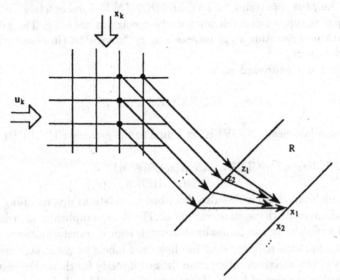

Fig. 1. Overview of the stucture of linguistic models (2)

The linguistic models expanded to cope with higher order dynamical properties can be described in a similar fashion. For instance, the expression,

$$x_{k+2} = (u_k \times x_k \times x_{k-1}) \circ R,$$

handles dynamical properties of the second order. Notice that the fuzzy relation R has a higher dimension.

Generally speaking, the above form of aggregation favors compatibility of the current state and the constraints (fuzzy relation) of the model.

The linguistic models can be also formulated in such a way that some other relationships between z_k and R are enhanced. We can enumerate two interesting dependencies:

— constraint inclusion. (2) is modified by combining z_k and R with the use of the relationship OR_INCLUDED:

$$x_{k+1} = OR\_INCLUDED(z_k, R) \tag{4}$$

In contrast to the previous relationship the level of membership of x_{k+1} becomes elevated once z_k becomes strongly included in the relevant model's constraints. The fuzzy model of this type is driven by constraint inclusion. Similarly as before (4) is converted coordinatewise and reads,

$x_{k+1}(x_i) = OR[z_k(z_1) \ INCLUDED\_INR(z_1, x_i),$

$z_k(z_2) \ INCLUDED\_IN \ R(z_2, x_i) \ldots]$

The two-argument predicate INCLUDED_IN can be realized by pseudocomplements ϕ induced by t-norms, $a\pi b = sup\{c \in [0,1] \mid atc \leq b\}$. If $a < b$ then $a\phi b = 1$ i.e., the constraint "$a \ INCLUDED\_IN \ b$" is completely fulfilled.

— The opposite type of dependency induces constraint covering. The grade of membership of resulting x_{k+1} increases as z_k "covers" R (in other words R is included in z_k).

The equation is expressed as

$$x_{k+1} = OR\_COVER (z_k, R)$$

where the relationship "COVER" is a dual to the predicate INCLUDED_IN, namely,

$x_{k+1} = OR[R(z_1, x_i) \ INCLUDED\_IN \ z_k(z_1), R(z_2, x_i)$

$INCLUDED\_IN \ z_k(z_2) \ldots]$

The linguistic models can be also described by relationships handling incremental changes of the system variables. This is accomplished at a level of physical variables or can be realized through logical transformations.

The first approach requires that the linguistic labels be associated with increments of the variables rather than defined directly for them. For instance, a standard repertoire of labels (fuzzy partition) could include,

$\delta = \{positive \ change, \ zero \ change, \ negative \ change\},$

etc., where each of the labels is defined as a fuzzy set over a space of changes of a given variable. This gives rise to the model which visualizes relationships between the changes,

$$\delta x_{k+1} = (\delta u_k \times \delta x_k) \circ T.$$

This class of models constitute a subsymbolic version of a class of relationships emerging in Qualitative Modelling[2, 1].

In the second approach, in addition to the linguistic labels of the frame, one introduces several landmarks (fuzzy sets) in the system variables around which the model is constructed. By expressing them in terms of the linguistic labels one obtains vectors s and t, respectively.

The formula of the model is put down as,

$$x_{k+1} \equiv t = (u_k \equiv s) \times (x_k \equiv t) \circ R$$

The model captures increments of the variables around s and t so that the structure handles relationships between incremental changes. The model becomes local and focussed on the dynamics occurring there.

The equivalence operator $\equiv$ (A EQUAL B) is implemented with the use of the previous INCLUDED_IN and COVER predicates (relations). It summarizes properties of both of them[5] namely,

A EQUAL B $\overset{\text{def}}{=}$ (A INCLUDED_IN B) AND (B INCLUDED_IN A).

Another definition includes complemented grades of membership A and B (i.e., $\overline{A}$ and $\overline{B}$).

$\overline{A}$ EQUAL $\overline{B}$ $\overset{\text{def}}{=}$ AVERAGE ((A INCLUDED_IN B) AND (B INCLUDED_IN A),

with a standard AVERAGE operation defined as,

$$\text{AVERAGE } (C, D) = (C + D)/2.0.$$

In comparison to the previous models, the reference model does not provide direct grades of membership of the elements of the partition but the levels of equality identified between x_{k+1} and the reference point t. Denoting the right hand side of the model by γ we get the equation,

$$x_{k+1} \equiv t \equiv \gamma,$$

which should be solved with respect to x_{k+1}. For γ lower than 1 (a vector with all entries equal to 1) the solution is not unique and induces regions of membership of states determined by γ. For further computational details refer to Pedrycz[5]. The lower the values of γ, the broader the range of membership of x_{k+1}, and the higher the uncertainty in determination of the values of the predicted state.

In a general setting the proposed linguistic models are viewed as single-level fuzzy relational equations[8]. The solutions methods of the families of equations, either exact or approximate, allow to address properly the emerging identification problem[3]. More general multilevel relational structures are available as well[6].

4 Conclusions

The aim of the paper was to discuss a role fuzzy sets as an interesting and computationally tractable conceptual platform for symbolical and numerical information processing. The role of linguistic labels has been presented. We also studied their contribution towards formation of a suitable cognitive perspective within which fuzzy dynamic models are constructed. Diverse facets of models involving fuzzy sets including conceptual as well as computational aspects have been explored. We carefully examined the nature of fuzzy modelling highlighting the levels of achieved generality. Various classes of specialized models have been introduced and analyzed.

Acknowledgement

Support from the Natural Sciences and Engineering Research Council of Canada and MICRONET is fully acknowledged.

References

1. D'Ambrosio, B.: Qualitative Process Theory Using Linguistic Variables. Springer-Verlag, New York, Berlin, (1989).
2. De Kleer, J., Brown, J.S.: A qualitative physics based on confluence. Artificial Intelligence, 24, 7-83, (1984).
3. Di Nola, A., Sessa, S., Pedrycz, W., Sanchez, S.: Fuzzy Relational Equations and Their Applications in Knowledge Engineering. Kluwer Academic Press, Dordrecht, (1989).
4. Menger, K.: Statistical metric spaces. Proceedings of the National Academy of Sciences. USA, 28, 535-537, (1942).
5. Pedrycz, W.: Direct and inverse problem in comparison of fuzzy data. Fuzzy Sets and Systems, 34, 223-236, (1990).
6. Pedrycz, W.: Fuzzy neural networks with reference neurons as pattern classifiers. IEEE Trans. Neural Networks, (1992)
7. Puccia, Ch.J. and Levins, R.: Qualitative Modeling of Complex Systems. Harvard University Press, Cambridge, MA, London, (1985).
8. Sanchez, E.: Resolution of composite fuzzy relation equations. Information and Control, 34, 38-48, (1976).
9. Zadeh, L.A.: Fuzzy sets. Information and Control, 8, 338-353, (1965).
10. Zadeh, L.A.: Outline of a new approach to the analysis of complex systems and decision processes. IEEE Trans. Syst., Man and Cybern, 1, 28-44, (1973).
11. Zadeh, L.A.: Fuzzy sets and information granularity. in: (M.M. Gupta, R.K. Ragade, R.R. Yager, eds.), Advances in Fuzzy Set theory and Applications, North Holland, Amsterdam, 3-18, (1979).

Handling Uncertainty, Context, Vague Predicates, and Partial Inconsistency in Possibilistic Logic

Didier Dubois, Jerome Lang and Henri Prade

Institut de Recherche en Informatique de Toulouse (IRIT), Université Paul Sabatier - CNRS, 118 route de Narbonne, 31062 Toulouse Cedex, France Tel. : (+33) 61.55.63.31 / 65.79 - Fax. : (+33) 61.55.62.39

Abstract. This short paper intends to provide an introduction to possibilistic logic, a logic with weighted formulas, to its various capabilities and to its potential applications. Possibilistic logic, initially proposed in [11], see also Léa Sombé [26] for an introduction, can be viewed as an important fragment of Zadeh[32]'s possibility distribution-based theory of approximate reasoning, put in a logical form. Possibilistic logic also relies on an ordering relation reflecting the relative certainty of the formulas in the knowledge base. As it will be seen, its semantics is based on a possibility distribution which is nothing but a convenient encoding of a preference relation a la Shoham[29], between interpretations. This kind of semantics should not be confused with the similarity relation-based semantics recently proposed by Ruspini[28] for fuzzy logics which rather extends the idea of interchangeable interpretations in a coarsened universe, e.g. Fariñas del Cerro and Orlowska[17], and which corresponds to another issue.

1 Necessity-weighted formulas and their semantics

Let us first recall that a necessity measure, denoted by N, is a function from a logical (propositional or first-order) language $\mathcal{L}$ to $[0, 1]$, such that $N(\top) = 1$ and $N(\bot) = 0$, where $\top$ (resp. $\bot$) denotes any tautology (resp. any contradiction), and obeying the following axiom

$$\forall p, \forall q, N(p \wedge q) = min(N(p), N(q)) \tag{1}$$

As a consequence we have $min(N(p), N(\neg p)) = 0$. However we only have $N(p \vee q) \geq max(N(p), N(q))$, since e.g. for $q = \neg p, N(p \vee \neg p) = N(\top) = 1$ while in case of total ignorance we may have $N(p) = N(\neg p) = 0$. A necessity measure N is the dual of a possibility measure Π, such that $\forall p, N(p) = 1 - \Pi(\neg p)$, where Π obeys the characteristic axiom $\forall p, \forall q, \Pi(p \vee q) = max(\Pi(p), \Pi(q))$ ([31]).

A possibilistic logic formula is a pair (p, α) where p is a classical first-order or propositional logic formula and α a number belonging to the semi-open real interval $(0, 1]$, which estimates to what extent it is certain that p is true considering the available information we have at our disposal. More formally (p, α) is a syntactic way to code the semantic constraint $N(p) \geq \alpha$.

As suggested by the operator min in (1), and since necessity measures, as well as possibility measures, are the perfect numerical counterparts of qualitative relations aiming at modelling "at least as certain" and "at least as possible" ([13]), the numbers used for weighting the formulas have an ordinal flavor. It departs from other uncertainty-handling approaches to automated reasoning, e.g. Baldwin[1], making use of probability bounds, and thus of sum and product operations.

N.B. : In this paper, for the sake of simplicity, we only consider lower bounds of necessity measures for weighting the formulas. We can also deal with lower bounds of possibility measures, i.e. $\Pi(p) \geq \alpha$, which corresponds to a very weak form of information; see [12, 23].

Let $M(p)$ be the set of models of p. The semantics of (p, α) is represented by the fuzzy set of models $M(p, \alpha)$ defined by [5] where ω denotes an interpretation

$$\mu_{M(p,\alpha)}(\omega) = 1 \text{ if } \omega \in M(p); \quad \mu_{M(p,\alpha)}(\omega) = 1 - \alpha \text{ if } \omega \notin M(p) \qquad (2)$$

In other words, the interpretations compatible with (p, α) are restricted by the above possibility distribution. The ones in $M(p)$ are considered as fully possible while the ones outside are all the more possible as α is smaller, i.e. the piece of knowledge is less certain. Note also that here we use the least specific possibility distribution $\pi = \mu_{M(p,\alpha)}$, i.e. the one with the greatest possibility degrees compatible with $\Pi(\neg p) \leq 1 - \alpha \Leftrightarrow N(p) \geq \alpha$, where the expression of a possibility measure Π in terms of a possibility distribution π is given by,

$$\forall p, \Pi(p) = sup\{\pi(\omega), \omega \in M(p)\}. \qquad (3)$$

(as a consequence of the characteristic axiom of possibility measures). Indeed, any possibility distribution π satisfying the constraint $N(p) \geq \alpha$ is such that $\forall \omega, \pi(\omega) \leq \mu_{M(p,\alpha)}(\omega)$.

In case of several pieces of knowledge $(p_i, \alpha_i), i = 1, \ldots, n$, forming a knowledge base $\mathcal{K}$, in agreement with the minimal specificity principle[14], we associate the following possibility distribution, built by performing the largest conjunction operation (which is also the only idempotent one) on the membership functions $\mu_{M(p_i, \alpha_i)}$, namely,

$$\pi_{\mathcal{K}}(\omega) = min_{i=1,\ldots,n} \mu_{M(p_i, \alpha_i)}(\omega). \qquad (4)$$

It can be checked that the necessity measure $N_{\mathcal{K}}$ induced from $\pi_{\mathcal{K}}$ by $N_{\mathcal{K}}(q) = 1 - sup\{\pi_{\mathcal{K}}(\omega), w \models \neg q\}$, where $w \models q$ means $\omega \in M(q)$, is the smallest necessity measure satisfying the constraints $N(p_i) \geq \alpha_i$. This is nothing but another formulation of the minimal specificity principle.

Let us take an example:

$$\mathcal{K} = \{(\neg p \vee q, 1), (\neg p \vee r, 0.7), (\neg q \vee r, 0.4), (p, 0.5), (q, 0.8)\}.$$

This induces the constraints,

$$\pi(\omega) \leq \mu_{M(\neg p \lor q, 1)}(\omega) \qquad \forall \omega \models p \land \neg q, \pi(\omega) = 0 \qquad sup\{\pi(\omega), \omega \models p \land \neg q\} = 0$$
$$\pi(\omega) \leq \mu_{M(\neg p \lor r, 0.7)}(\omega) \qquad \forall \omega \models p \land \neg r, \pi(\omega) \leq 0.3 \qquad sup\{\pi(\omega), \omega \models p \land \neg r\} \leq 0.3$$
$$\pi(\omega) \leq \mu_{M(\neg q \lor r, 0.4)}(\omega) \Leftrightarrow \forall \omega \models q \land \neg r, \pi(\omega) \leq 0.6 \Leftrightarrow sup\{\pi(\omega), \omega \models q \land \neg r\} \leq 0.6$$
$$\pi(\omega) \leq \mu_{M(p, 0.5)}(\omega) \qquad \forall \omega \models \neg p, \pi(\omega) \leq 0.5 \qquad sup\{\pi(\omega), \omega \models \neg p\} \leq 0.5$$
$$\pi(\omega) \leq \mu_{M(q, 0.8)}(\omega) \qquad \forall \omega \models \neg q, \pi(\omega) \leq 0.2 \qquad sup\{\pi(\omega), \omega \models \neg q\} \leq 0.2$$

and the corresponding possibility distribution $\pi_{\mathcal{K}}$, defined by (3), is pictured on Fig 1.

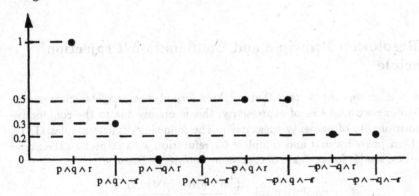

Fig. 1

The possibility distribution of Fig. 1 is normalized, i.e. $\exists \omega, \pi_{\mathcal{K}}(\omega) = 1$; in the general case, there may exist several ω such that $\pi_{\mathcal{K}}(\omega) = 1$. This means that $\mathcal{K}$ is fully consistent since there is at least one interpretation in agreement with $\mathcal{K}$ which is completely possible. More generally we define the degree of inconsistency of $\mathcal{K}$ by,

$$Inc(\mathcal{K}) = 1 - sup_{\omega} \pi_{\mathcal{K}}(\omega) \qquad (5)$$

It can also be established that $Inc(\mathcal{K}) > 0 \Leftrightarrow \mathcal{K}*$ is inconsistent, where $\mathcal{K}*$ is the classical knowledge base obtained from $\mathcal{K}$ by deleting the weights.

A possibility distribution such as the one pictured in Fig. 1 is a way of encoding a preference ordering among interpretations, i.e. the kind of relation used by Shoham[29] for providing non-monotonic logics with a semantics. Indeed, it has been shown[15] that the preferential entailment $\models_{\pi}$, where π is short for $\pi_{\mathcal{K}}$, defined by,

$$p \models_{\pi} q \Leftrightarrow (\exists \omega, \omega \models_{\pi} p \text{ and } \forall \omega, \omega \models_{\pi} p \Rightarrow \omega \models q),$$

where

$$\omega \models_{\pi} p \Leftrightarrow \omega \models p, \Pi(p) > 0 \text{ and } \not\exists \omega', \omega' \models p \text{ and } \pi(\omega) < \pi(\omega'),$$

is in complete agreement with non-monotonic consequence relations obeying the axiomatics of system P proposed by Kraus et al.[21]. It can be also shown that it is closely related to the notion of conditional possibility since we have the equivalences,

$$p \models_\pi q \Leftrightarrow \Pi(q|p) > \Pi(\neg q|p) \text{ with } \Pi(q|p) = \begin{cases} 1 & \text{if } \Pi(p) = \Pi(p \wedge q) \\ \Pi(p \wedge q) & \text{if } \Pi(p) > \Pi(p \wedge q) \end{cases}$$

$$\Leftrightarrow N(q|p) > 0 \qquad \text{with } N(q|p) = 1 - \Pi(\neg q|p)$$

$p \models_\pi q$ means that preferred models of p (as induced by π) are all models of q.

2 Resolution Principle and Combination/Projection Principle

In this section we suppose that that weighted formulas are weighted clauses; this can be done without loss of expressivity; this is mainly due to the conjunctive compositionality of necessity measure[7]. The following deduction rules[11, 12] have been proved sound and complete for refutation with respect to the above semantics; see [5] for the case of consistent knowledge bases

$$\text{resolution rule: } \frac{(\neg p \vee q, \alpha) \ (p \vee r, \beta)}{(q \vee r, min(\alpha, \beta))}$$

$$\text{particularization: } \frac{(\forall x \ p(x), \alpha)}{(p(a), \alpha)} \text{ (as well as more general substitutions)}$$

If we want to compute the maximal certainty degree which can be attached to a formula according to the constraints expressed by a knowledge base $\mathcal{K}$, for instance r in the above example, we add to $\mathcal{K}$ the clause(s) obtained by refuting the proposition to evaluate with a necessity degree equal to 1, here we add $(\neg r, 1)$. Then it can be shown that any lower bound obtained on $\perp$, by resolution, is a lower bound of the necessity of the proposition to evaluate. See [4] for an ordered search method which guarantees the obtaining of the greatest derivable lower bound on $\perp$. It can be shown[5, 23], that this greatest derivable lower bound on $\perp$ is nothing but $Inc(\mathcal{K} \cup \{(\neg r, 1)\})$ where r is the proposition to establish. In the example, we have the following derivation,

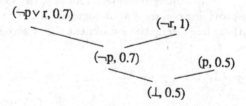

i.e. $N(r) \geq 0.5$ and indeed it can be checked that using the possibility distribution pictured in Fig. 1, we have $\Pi(\neg r) = sup\{\pi(\omega), \omega \models \neg r\} = 0.5$, in fact $\Pi(\neg r) \leq 0.5$ since π is the greatest possibility distribution compatible with $\mathcal{K}$, and then $N(r) = 1 - \Pi(\neg r) \geq 0.5$.

It is has been pointed out in [7] that this procedure is in agreement with Zadeh's approach to approximate reasoning based on combination/projection of possibility distributions. This can be also checked on our example. Let us suppose that p means "$X \geq a$", q means "$X \geq b$" with $b < a$ and r means "$Y \in [c, d]$" where X and Y are two real-valued variables under consideration. Then it can be seen that the 8 interpretations correspond to,

$$S(p \wedge q \wedge r) = S(p) \cap S(q) \cap S(r) = [a, +\infty) \times [c, d];$$
$$S(p \wedge q \wedge \neg r) = [a, +\infty) \times \overline{[c, d]};$$
$$S(p \ \wedge \neg q \wedge r) = \emptyset; \quad S(p \wedge \neg q \wedge \neg r) = \emptyset;$$
$$S(\neg p \wedge q \wedge r) = [b, a) \times [c, d]; \quad S(\neg p \wedge q \wedge \neg r) = [b, a) \times \overline{[c, d]};$$
$$S(\neg p \wedge \neg q \wedge r) = [0, b) \times [c, d]$$
$$S(\neg p \wedge \neg q \wedge \neg r) = [0, b) \times \overline{[c, d]}.$$

where $S(\omega)$ denotes the set of values of the pair (X, Y) corresponding to the interpretation ω. Then $\mathcal{K}$ is equivalent to a set of possibility distributions, each one corresponding to a piece of knowledge, namely,

$(\neg p \vee q, 1) : \{\pi^1_{X,Y}(u, v) = 1, \ \forall(u, v)$ because $S(p \wedge \neg q) = \emptyset$, i.e. $\neg p \vee q$ is a tautology

$$(\neg p \vee q, 0.7) : \pi^2_{X,Y}(u, v) \begin{cases} \leq 0.3, \forall(u, v) \in [b + \infty) \times \overline{[c, d]} \\ = 1 \quad \text{otherwise;} \end{cases}$$

$$(\neg q \vee r, 0.4) : \pi^3_{X,Y}(u, v) \begin{cases} \leq 0.6, \forall(u, v) \in [a, +\infty) \times \overline{[c, d]} \\ = 1 \quad \text{otherwise;} \end{cases}$$

$$(p, 0.5) : \pi^4_{X,Y}(u, v) \begin{cases} = 1 \quad \text{if } u \in [a, +\infty) \\ \leq 0.5 \text{ otherwise;} \end{cases}$$

$$(q, 0.8) : \pi^5_{X,Y}(u, v) \begin{cases} = 1 \quad \text{if } u \in [b, +\infty) \\ \leq 0.2 \text{ otherwise.} \end{cases}$$

Then it can be checked that,
$$\pi_Y(v) = \sup_u min(\pi^1_{X,Y}(u, v), \pi^2_{X,Y}(u, v), \pi^3_{X,Y}(u, v), \pi^4_{X,Y}(u, v), \pi^5_{X,Y}(u, v)) =_{def} \sup_u \pi^*_{X,Y}(u, v)$$
with

$$\pi^*_{X,Y}(u, v) \begin{cases} 1 \quad \text{if } (u, v) \in [a, +\infty) \times [c, d] \\ \leq 0.5 \text{ if } u \in [b, a) \end{cases}$$

$$\pi^*_{X,Y}(u, v) \begin{cases} \leq 0.3 \text{ if } (u, v) \in [a, +\infty) \times \overline{[c, d]} \\ \leq 0.2 \text{ if } u \in \overline{[b, +\infty)} \end{cases}$$

Then $\pi_Y(v) = 1$ if $v \in [c, d]$; $\pi_Y(v) \leq 0.5$ if $v \notin [c, d]$ and finally $\Pi(S(\neg r)) = \Pi(\overline{[c, d]}) = \sup_{v \in \overline{[c, d]}} \pi_Y(v) \leq 0.5$ and thus $N(S(r)) \geq 0.5$, i.e. we recover the result already obtained by refutation.

N.B.: Semantic evaluation methods, extending a procedure by Davis and Putnam, are also available in possibilistic logic[22].

3   Labelled Formulas and the Handling of Vague Predicates

As pointed out in [5], the weighted clause $(\neg p \vee q, \alpha)$ is semantically equivalent to the weighted clause $(q, min(\alpha, v(p)))$ where $v(p)$ is the truth value of p, i.e. $v(p) = 1$ if p is true and $v(p) = 0$ if p is false. Indeed, for any uncertain proposition (p, α) we can write $\mu_{M(p,a)}(\omega)$ under the form $max(v_\omega(p), 1 - \alpha)$, where $v_\omega(p)$ is the truth-value assigned to p by interpretation ω. Then obviously:

$$\forall \omega, \mu_{M(\neg p \vee q, \alpha)}(\omega) = max(v_\omega(\neg p \vee q), 1 - \alpha)$$
$$= max(1 - v_\omega(p), v_\omega(q), 1 - \alpha)$$
$$= max(v_\omega(q), 1 - min(v_\omega(p), \alpha))$$
$$= \mu_{M(q, min(v_\omega(p), \alpha))}(\omega).$$

This remark is very useful for hypothetical reasoning, since by "transferring" an atom from a clause to the weight part of the formula we are introducing explicit assumptions. Indeed changing $(\neg p \vee q, \alpha)$ into $(q, min(v(p), \alpha))$ leads to state the piece of knowledge under the form "q is certain at the degree α, *provided that* p is true". Then the weight is no more just a degree but in fact a label which expresses the context in which the piece of knowledge is more or less certain.

More generally, the weight or label can be a function of logical (universally quantified) variables involved in the clause. Thus a possibilistic formula of the form $(c(x), \mu_P(x))$ expresses that for any x, one is certain that the clause $c(x)$ is true with a necessity degree greater or equal to $\mu_P(x)$ where μ_P is the membership function of a predicate. This predicate can be an ordinary predicate or a fuzzy predicate. In this latter case the possibilistic formula means "the larger $\mu_P(x)$, the more certain $c(x)$". Note that in any case the clause remains a classical clause, while the fuzzy predicate appears in the weight. For instance the rule "the younger the person, the more certain he/she is single" will be represented by $(single(x), \mu_{young}(age(x))$ where young has a membership function which has the value 1 until the legal age for marriage and then decreases. With such a clause, once instantiated, with, say $x =$ John, we need to know the age of John for computing the certainty degree. If we only have a fuzzy knowledge about John's age, modelled by a possibility distribution π, we have to change $\mu_{young}(age(John))$ which is not known, by $N_\pi(young) = inf_u \, max(\mu_{young}(u), 1 - \pi(u))$, i.e. the certainty that John is indeed definitely young given the available information about his age.

Fuzzy pieces of knowledge like "John is young" can be also modelled in possibilistic logic. Let us assume for "young" a membership function as the one described above. Then the piece of knowledge is equivalent to the family of weighted formulas making use of the non-fuzzy predicates "$\leq (age(x), a)$" expressing that $age(x) \leq a$, where the logical constant "a" ranges on the domain of attribute 'age'. The degree of necessity of "$\leq (age(John), a)$" given that "John is young" is given by,

$$N_{young}(\leq (age(John), a)) = inf_{a < u} \, 1 - \mu_{young}(u)$$

which leads to the family of possibilistic formulas,

$$(\leq (age(John), a), N_{young}(\leq (age(John), a))).$$

Similarly "John is not young" will be represented by,

$$(> (age(John), a), N_{not\_young}(> (age(John), a))).$$

Note that the resolution of these two weighted formulas leads to $(\perp, sup_a\text{'}min(N_{young}(\leq (age(John), a)), N_{not\_young}(> (age(John), a)))$ which is equal to $1/2$ if μ_{young} is a continuous membership function and would be equal to 1 if 'young' were modelled in a crisp way by an interval. This is natural since the two pieces of knowledge we start with are fully contradictory only if we have a crisp understanding of the idea of "young".

Using a similar (but slightly different) interpretation of fuzzy predicates, the resolution rule has been extended in [12], to the case of weighted fuzzy propositions (which thus no longer belong to a Boolean algebra). In that case it is possible to explicitly deal with clauses like $(\neg young \vee single, \alpha)$.

4 Coping with Inconsistency

A nice feature of possibilistic logic is its capacity to cope with a partially inconsistent knowledge base $\mathcal{K}$ such that $Inc(\mathcal{K}) > 0$. Roughly speaking, the conclusions which can be obtained with a degree of uncertainty strictly higher than $Inc(\mathcal{K})$ are still meaningful since for sure only a consistent subpart of $\mathcal{K}$ (containing the most certain pieces of knowledge) is used for deducing them. Indeed in any inconsistent sub-base of $\mathcal{K}$ there is (at least) a clause with a weight less or equal to $Inc(\text{K})$.

For instance, let us consider the knowledge base $\mathcal{K}$ previously introduced to which we add the clause $(\neg p \vee \neg q, 0.2)$; let $\mathcal{K}' = \mathcal{K} \cup \{(\neg p \vee \neg q, 0.2)\}$. Clearly we have $\pi_{\mathcal{K}'}(p \wedge q \wedge r) = 0.8$ now (instead of $\pi_{\mathcal{K}}(p \wedge q \wedge r) = 1$ in Fig. 1), and $Inc(\mathcal{K}') = 0.2$. A minimal inconsistent sub-base of $\mathcal{K}'$ is $\{(\neg p \vee \neg q, 0.2), (p, 0.5), (q, 0.8)\}$. From $\mathcal{K}'$ we can still deduce $(r, 0.5)$ or $(p, 0.5)$ using only consistent parts of $\mathcal{K}'$ since $0.5 > Inc(\mathcal{K}')$. Suppose now that we add $(\neg p \vee \neg q, 0.6)$ instead of $(\neg p \vee \neg q, 0.2)$. Let $\mathcal{K}'' = \mathcal{K} \cup \{(\neg p \vee \neg q, 0.6)\}$. Then $Inc(\mathcal{K}'') = 0.5$, and now we can deduce $(\neg p, 0.6)$ by refutation from $\mathcal{K}''$ using the consistent part of $\mathcal{K}'' \cup \{(\neg p \vee \neg q, 0.6), (q, 0.8)\}$, since $0.6 > Inc(\mathcal{K}'')$ while $(p, 0.5)$ is no longer considered as an allowed deduction since $0.5 = Inc(\mathcal{K}'')$. Thus a non-monotonic reasoning process is at work in possibilistic logic when partial inconsistency is introduced in the knowledge base. In Lang et al.[23] an inconsistency-tolerant semantics is proposed, adding an "absurd interpretation" on which the possibility distribution attached to the knowledge base is normalized. Using this semantics the soundness and completeness results for refutation that we have in the consistent case can still be shown to hold. Moreover it has been established in [15] that we have $N_{\mathcal{K}}(q|p) > 0$ if and only if it is possible to deduce (q, β) from $\mathcal{K} \cup \{(p, 1)\}$ with $\beta > Inc(\mathcal{K} \cup \{(p, 1)\})$, where $N_{\mathcal{K}}$ is the necessity measure defined from $\pi_{\mathcal{K}}$. If we define $p \hspace{-0.3em}\sim q$ by $N_{\mathcal{K}}(q|p) > 0$, then $\hspace{-0.3em}\sim$ is a non-monotonic consequence relation [15]; this illustrates the close relation that exists between non-monotonic reasoning and belief revision [27], in the possibilistic framework. The reader is referred to [16] for a detailed analysis of belief revision in possibility theory. Besides, the problem of recovering consistency in a partially inconsistent knowledge base $\mathcal{K}$ by building maximal consistent sub-bases (obtained by deleting suitable pieces of knowledge in $\mathcal{K}$) is discussed in [9]. Note that such a syntactic approach is not necessarily equivalent to a treatment based on a semantic representation such as $\pi_{\mathcal{K}}$, since syntactically distinct knowledge bases $\mathcal{K}$ and $\mathcal{K}'$ may be such that $\pi_{\mathcal{K}} = \pi_{\mathcal{K}'}$.

Lastly, another way of dealing with inconsistency might be to allow for paraconsistent pieces of knowledge. Roughly speaking, the idea of paraconsistency, first introduced by da Costa[3], is to say that we have a paraconsistent knowledge about p if we both want to state p and to state $\neg p$. It corresponds to the situation where we have conflicting information about p. In a paraconsistent logic we do no want to have every formula q deducible as soon as the knowledge base contains p and $\neg p$ (as it is the case in classical logic). The idea of paraconsistency is "local" by contrast with the usual view of inconsistency which considers the knowledge base in a global way. In possibilistic logic it would correspond to have p in $\mathcal{K}$ with both $N(p) \geq \alpha > 0$ and $N(\neg p) \geq \beta > 0$. The situation may be perhaps better understood if we consider Zadeh's combination/projection point of view first. Indeed let us suppose that, for some variable X (representing the value of some attribute), our knowledge is represented by a *non-normalized* possibility distribution π_X. Then $\forall A \subseteq domain(X)$, we have $min(N_X(A), N_X(\overline{A})) > 0$ where the necessity measure N_X is based on π_X. When combining this piece of knowledge π_X with other possibility distributions π^i, it can be easily seen that the resulting possibility distribution π will be such that its height $h(\pi) =_{def} sup_u \pi(u) = h(\pi_X)$ iff $h(\pi_i) \geq h(\pi_X)$, $\forall i$. In other words, *if used in a reasoning process with other non-paraconsistent pieces of knowledge,* π_X will affect the result by "denormalizing" the resulting possibility distribution, thus leading to a paraconsistent conclusion. This could be handled in a syntactic way in possibilistic logic as suggested by the following simple example of the modus ponens,

$$N(\neg p \vee q) \geq \alpha > 0$$
$$\frac{N(p) \geq \beta > 0; \quad N(\neg p) \geq \gamma > 0}{N(q) \geq min(\beta, max(\alpha, \gamma))} \; ; N(\neg q) \geq min(\beta, \gamma).$$

Indeed, we have $(\neg p \vee q) \wedge p \models q$, so $N(q) \geq N((\neg p \vee q) \wedge p) = min(N(\neg p \vee q), N(p)) \geq min(\alpha, \beta)$, and also $p \wedge \neg p \models q$, so $N(q) \geq min(N(p), N(\neg p)) \geq min(\beta, \gamma)$. Then $N(q) \geq max(min(\alpha, \beta), min(\beta, \gamma))$. The situation is pictured in Fig. 2

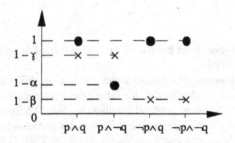

Fig. 2

It can be checked that $min(N(p), N(\neg p)) = min(N(q), N(\neg q)) \geq min(\beta, \gamma) > 0$, which expresses that the degree of paraconsistency is propagated to the conclusion. If $\gamma = 0$, a particular case of the resolution principle is recovered. This suggests the following way of dealing with paraconsistent knowledge: i) keep the paraconsistent pairs $N(p) \geq \alpha, N(\neg p) \geq \beta$ (with degree of paraconsistency $min(\alpha, \beta)$) separate from the remaining part of the knowledge base which is supposed to be consistent; ii) use the paraconsistent knowledge only, when it is impossible to produce a strictly positive lower bound on the necessity degree of a proposition of interest by using the consistent part of the knowledge base only. Then any conclusion which will be produced will have a degree of paraconsistency equal to the maximum of the degrees of paraconsistency of the paraconsistent pairs involved in the production of this conclusion. The idea to add paraconsistent knowledge only when necessary, to the consistent part of the knowledge base is a bit similar to the building of maximal consistent sub-bases[9]. The development of these ideas is left for further research.

5 Concluding Remarks: Potential Applications

First steps towards possibilistic logic programming can be found in [8]. The use of possibilistic logic as a programming language is all the more of interest that min-max discrete optimization problems (and more generally systems of possibly incompatible, prioritized constraints) can be expressed (and then solved) in possibilistic logic[23]. Applications to hypothetical reasoning for diagnosis purposes or for finding "optimal" maximal consistent sub-bases of an inconsistent possibilistic knowledge base are discussed in [6, 10] where possibilistic Assumption-based Truth Maintenance Systems are developed. Other developments of ideas very close to possibilistic logic can be found in Froidevaux and Grossetête[18] and in Chatalic and Froidevaux[2]. See also Jackson[20] for the computation of possibilistic prime implicants and their use in abduction. Besides these, Larsen and Nonfjall[25], Yager and Larsen[30] have used possibilistic logic in validation of knowledge bases.

References

1. Baldwin J.F.: Support logic programming. International Journal of Intelligent Systems 1:73-104, (1986).
2. Chatalic P., Froidevaux, C.: Graded logics : a framework for uncertain and defeasible knowledge. in Methodologies for Intelligent Systems (eds) Z.W. Ras, M. Zemankova, Lecture Notes in Artificial Intelligence, Vol. 542, Springer Verlag, (1991).
3. Da Costa N.C.A.: Calculus propositionnels pour les systèmes formels inconsistants. Compte Rendu Acad. des Sciences (Paris), 257:3790-3792, (1963).
4. Dubois D., Lang. J, Prade, H.: Theorem-proving under uncertainty A possibilistic theory-based approach. Proceedings of 10th International Joint Conference on Artificial Intelligence, Milano, Italy, 984-986, (1987).
5. Dubois D., Lang, J., Prade, H.: Automated reasoning using possibilistic logic: semantics, belief revision, variable certainty weights. Proceedings 5th Workshop on Uncertainty in AI, Windsor, Ont., 81-87, (1989).
6. Dubois D., Lang, J., Prade, H.: Handling uncertain knowledge in an ATMS using possibilistic logic. in: Methodologies for Intelligent Systems 5 (eds) Z.W. Ras, M. Zemankova, M.L. Emrich, North-Holland, Amsterdam, 252-259, (1990).
7. Dubois D., Lang, J., Prade, H.: Fuzzy sets in approximate reasoning - Part 2: Logical approaches. Fuzzy Sets and Systems, 25th Anniversary Memorial Volume, 40:203-244, (1991).
8. Dubois D., Lang, J., Prade, H.: Towards possibilistic logic programming. Proceedings 8th International. Conf. on Logic Programming (ICLP'91), Paris, June 25-28, MIT Press, Cambridge, Mass., (1991).
9. Dubois D., Lang, J., Prade, H.: Inconsistency in possibilistic knowledge bases - To live or not live with it. in: Fuzzy Logic for the Management of Uncertainty (L.A. Zadeh, J. Kacprzyk, eds.), Wiley, pp 335-351, (1992).
10. Dubois D., Lang, J., Prade, H.: A possibilistic assumption-based truth maintenance system with uncertain justifications, and its application to belief revision. Proceedings ECAI Workshop on Truth-Maintenance Systems, Stockholm, Lecture Notes in Computer Sciences, no. 515, Springer Verlag (J.P. Martins, M. Reinfrank, eds.), 87-106, (1991).
11. Dubois D., Prade. H.: Necessity measures and the resolution principle. IEEE Trans.on Systems, Man and Cybernetics 17:474-478, (1987).
12. Dubois D., Prade. H.: Resolution principles in possibilistic logic. International Journal of Approximate Reasoning 4(1): 1-21, (1990).
13. Dubois D., Prade. H:: Epistemic entrenchment and possibilistic logic. in: Tech. Report IRIT/90-2/R, IRIT, Univ. P. Sabatier, Toulouse, France, (1990). Artificial Intelligence, (50) pp 223-239, 1991.
14. Dubois D., Prade. H.: Fuzzy sets in approximate reasoning - Part 1 : Inference with possibility distributions. Fuzzy Sets and Systems, 25th Anniversary Memorial Volume 40:143-202, (1991).
15. Dubois D., Prade. H.: Possibilistic logic, preference models, non-monotonicity and related issues. Proceedings of the 12th International Joint Conference on Artificial Intelligence, Sydney, Australia, Aug. 24-30,pp 419-424, (1991).
16. Dubois D., Prade. H.: Belief revision and possibility theory. in : Belief Revision (ed) P. Gärdenfors, Cambridge University Press, pp 142-182, (1992).
17. Fariñas del Cerro L., Orlowska, E.: DAL-A logic for data analysis. Theoretical Comp. Sci. 36: 251-264, (1985).

18. Froidevaux C., Grossetète. C.: Graded default theories for uncertainty. Proceedings of the 9th European Conference on Artificial Intelligence (ECAI-90), 6-10, (1990), 283-288.

19. Gärdenfors P.: Knowledge in Flux - Modeling the Dynamics of Epistemic States. The MIT Press, Cambridge, (1988).

20. Jackson P.: Possibilistic prime implicates and their use in abduction. Research Note, McDonnell Douglas Research Lab., St Louis, MO, (1991).

21. Kraus S., Lehmann, D., Magidor, M.: Nonmonotonic reasoning, preferential models and cumulative logics. Artificial Intelligence 44(1-2):134-207, (1990).

22. Lang J.: Semantic evaluation in possibilistic logic. in: Uncertainty in Knowledge Bases (B. Bouchon-Meunier, R.R. Yager, L.A. Zadeh, eds.), Lecture Notes in Computer Science, Vol. 521, Springer Verlag, 260-268, (1991).

23. Lang J.: Possibilistic logic as a logical framework for min-max discrete optimisation problems and prioritized constraints. in: Fundamentals of Artificial Intelligence Research (P. Jorrand, J. Kelemen, eds.), Lecture Notes in Artificial Intelligence, Vol. 535, Springer Verlag, 112-126, (1991).

24. Lang J., Dubois. D., Prade, H.: A logic of graded possibility and certainty coping with partial inconsistency. Proceedings 7th Conference on Uncertainty in AI, Morgan Kaufmann, (1990).

25. Larsen H.L., Nonfjall, H.: Modeling in the design of a KBS validation system. Proceedings 3rd International Fuzzy Systems Assoc. (IFSA), (1989), 341-344.

26. Lèa Sombé, Besnard, P., Cordier, M., Dubois, D., Fariñas, C., del Cerro, Froidevaux, C.,Moinard, Y., Prade, H. Schwind, L., Siegel, D.: Reasoning Under Incomplete Information in Artificial Intelligence: A Comparison of Formalisms Using a Single Example. Wiley, New York, (1990).

27. Makinson D., Gärdenfors, P.: Relation between the logic of theory change and nonmonotonic logic. in: The Logic of Theory Change (eds) A. Fuhrmann, M. Morreau, Lecture Notes in Computer Sciences, no. 465, Springer Verlag, 185-205, (1991).

28. Ruspini E.H.: On the semantics of fuzzy logic. International Journal of Approximate Reasoning 5:45-88, (1991).

29. Shoham Y.: Reasoning About Change - Time and Causation from the Standpoint of Artificial Intelligence. MIT Press, Cambridge, Mass., (1988).

30. Yager, R.R., Larsen, H.L.: On discovering potential inconsistencies in validating uncertain knowledge bases by reflecting on the input. Tech. Rep. #MII-1001, Iona College, New Rochelle, N.Y., (1990).

31. Zadeh L.A.: Fuzzy sets as a basis for a theory of possibility. Fuzzy Sets and Systems 1:3-28, (1978).

32. Zadeh L.A.: A theory of approximate reasoning. in: Machine Intelligence 9 (eds) J.E. Hayes, D. Michie, L.I. Mikulich, Elsevier, New York, 149-194, (1979).

An Adaptive Fuzzy Model-based Controller

Bruce Graham and Robert Newell

Computer Aided Process Engineering Centre, Dept. of Chemical Engineering, The University of Queensland, 4072, Australia, Email: bobn@cape.uq.oz.au

Abstract. A method of identifying a process model from plant input-output data has been developed. The model is in the form of qualitative linguistic relationships which are represented and evaluated using fuzzy set theory[7]. This so-called fuzzy identification is used in the design of fuzzy model-based controllers. On-line identification is used to produce an adaptive fuzzy controller.

1 Introduction

For nearly twenty years automatic controllers based on fuzzy set theory have been developed[6]. The most common form of fuzzy controller uses a set of heuristic "rules of thumb", and is called a fuzzy heuristic controller. An alternative fuzzy control algorithm has been proposed by Hendy[5] - the fuzzy model-based controller. This controller uses linguistic information in the form of a "cause and effect process model.

Due to their use of qualitative linguistic information, fuzzy controllers are not readily amenable to a rigorous mathematical analysis. Consequently, little has been achieved in formulating rigorous design techniques for fuzzy controllers.

Our work formalises the design of the fuzzy model-based controller through the use of fuzzy identification of the process model. Fuzzy identification of the process model directly from process input-output data removes the reliance upon sometimes unreliable human expertise from the design step. It also enables the design of a controller for a process about which even elementary heuristic control rules are not known.

This work also develops an adaptive fuzzy controller by modifying the structure of the model-based controller by adding on-line identification of the fuzzy process model[3, 4, 1]. A schematic of the controller is shown in Fig. .1.

This controller can be used in different ways to solve a number of problems:

1. to self-tune a model-based controller given a crude initial model - this enables a simple model developed a priori to be matched to the process.
2. to adapt to changing process conditions by modifying the process model - this means only a small model pertaining to the current operating point is required, rather than a large model that describes the process across a broad operating range.
3. to learn a control strategy by the identification of the process model, starting with no model at all - this fully automates the design of a fuzzy model-based controller.

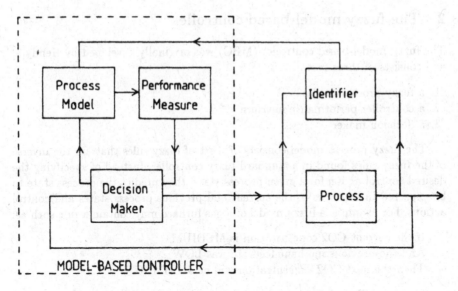

Fig. 1 The adaptive fuzzy controller

A study of fuzzy identification algorithms has been carried out and the problems of model completeness and model consistency have been dealt with. Five case studies illustrate and prove the algorithms developed. The case studies are:

1. silica levels in alumina from the Bayer process
2. a mass storage tank
3. a cement kiln
4. a liquid level rig
5. a single effect forced circulation evaporator

The first of these is an industrial study conducted at ALCOA's Kwinana refinery. The liquid level rig is a laboratory-scale rig. The remaining studies are computer simulations. These studies show that, given adequate process data, the identification of a fuzzy process model, and hence the design of a fuzzy model-based controller, follows easily.

The last four case studies demonstrate the use of the adaptive fuzzy controller in each of its possible modes given above. It is shown to successfully self-tune, adapt and learn a control strategy. This controller structure significantly enhances the design and use of the fuzzy model-based controller.

This work was carried out while the first author was completing his PhD thesis in the Department of Chemical Engineering at the University of Queensland, Australia.

2   The fuzzy model-based controller

The fuzzy model-based controller (MBC) was originally developed by Hendy[5] and consists of three parts:

1. a fuzzy process model
2. a controller performance measure
3. a decision maker

The fuzzy process model consists of a set of fuzzy rules that are the inverse of the fuzzy rules found in a standard fuzzy controller. Instead of specifying the desired control action for a given process state, they predict the process state to be expected in time due to the current and previous process states and control actions. For example, a fuzzy model of a gas furnace may contain a rule such as:

If the current CO_2 concentration is MEDIUM
And the previous methane feedrate was LOW
Then the next CO_2 concentration will be **JUST HIGH**

The controller performance measure consists of a collection of fuzzy sets. Each fuzzy set describes the required performance of a process variable. The performance of a particular variable is given by the degree of membership of its current value in its performance fuzzy set. Two examples of typical performance fuzzy sets are shown in Fig. .2.

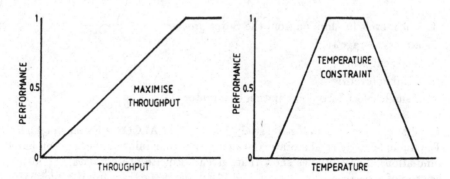

Fig. 2 Two performance fuzzy sets: (a) a fuzzy goal (b) a fuzzy constraint.

The overall controller performance is given by a confluence rule that combines the individual performance values, for example:

Overall performance = Throughput performance AND Temperature performance

The "AND" connective might be implemented as the minimum or the product of the individual values to enable the calculation of a single performance value.

The controller output is calculated by the decision maker, using the fuzzy process model and the performance measure. The aim of the decision maker is to choose, from a predefined, finite set of control actions, which action will maximise the performance of the process if no future control actions are taken. This is achieved by applying each possible control action to the fuzzy process model to obtain predictions of the process state to be expected due to each control action. The control action that produces the predicted process state of highest performance, as specified by the performance measure, is chosen as the current controller output.

3 Fuzzy model identification

The design of the MBC can be aided by the use of fuzzy identification of the process model from process input-output data. This alleviates the need to specify the model from expert knowledge of the process, which may not be available, or at least may be unreliable.

In relation matrix terminology, the fuzzy model of the process is of the form:

$$X = U^1 \circ U^2 \circ \ldots \circ U^N \circ R \tag{1}$$

where X is the predicted fuzzy state of the process, the U^i are the current and previous fuzzy values of the process state and controller output, and R is the fuzzy relation. The aim of the fuzzy identification is to identify the relation matrix, R, from a given set of input-output data,

$$[X_k; U_k^1, U_k^2, \ldots, U_k^N]_{k=1}^K \tag{2}$$

Each set of input-output data defines a relation matrix, R_k, satisfying the relation

$$X_k = U_k^1 \circ U_k^2 \circ \ldots \circ U_k^N \circ R_k \tag{3}$$

There is a family of relation matrices that satisfy this relation. One easily calculated solution is given by:

$$R_k = U_k^1 \times U_k^2 \times \ldots \times U_k^N \times X_k \tag{4}$$

where $\times$ is the Cartesian product. The overall relation matrix, R, derived from the entire set of data, is calculated by:

$$R = \cup_{k=1}^K R_k \tag{5}$$

If the data set is such that

$$X_k = U_k^1 \circ U_k^2 \circ \ldots \circ U_k^N \circ \tilde{R} \qquad \forall k = 1, 2, \ldots, K \tag{6}$$

for some relation matrix, $\tilde{R}$, then, given particular conditions on the data set,

$$R = \tilde{R} \tag{7}$$

In general, though, this will not be true as data collected from the process is likely to be noisy and incomplete. In this case the relation matrix, R, forms only an approximate fuzzy model of the process.

4 Controller adaptation via on-line identification

Adaptation of the controller is achieved by the use of on-line fuzzy identification of the process model. The controller may start with no model at all, or a predefined model. The on-line identification will build a model, or modify the predefined model to more closely match the process. As the identification proceeds, the performance of the controller improves.

The adaptive fuzzy model-based controller (Fig. .1) uses on-line fuzzy identification in the following algorithm to obtain controller adaptation:
At time step 0:

1. An intial relation matrix, R_0, is defined
2. The initial model inputs, $U_0^1, \ldots, U_0^N$ are established

At time step k+1:

1. The relation matrix model is updated using fuzzy identification via

$$R_{k+1} = R_k \cup (U_k^1 \times U_k^2 \times \ldots \times U_k^N \times X_{k+1}) \qquad (8)$$

2. The model-based controller calculates a new controller output using the fuzzy model

$$X = U^1 \circ U^2 \circ \ldots \circ U^N \circ R_{k+1} \qquad (9)$$

3. The controller output is

$$[U_{k+1}^1, U_{k+1}^2, \ldots, U_{k+1}^N] \qquad (10)$$

The initial relation matrix may be completely empty i.e. $R_0 = 0$, or it may be predefined using, say, identification from process data, as described above. The model that is identified on-line is a predictive model that predicts the expected process state over one sample time.

A problem with the fuzzy model, especially if it is initially empty, is that it may be incomplete during the initial stages of on-line identification. This means that it cannot give a prediction for the expected process state from the current set of model inputs. In this case the model-based controller cannot calculate a controller output and so an auxiliary algorithm must be used to decide on an appropriate control action. This algorithm may be extremely simple, such as to leave the controller output fixed until such time as the model-based controller is able to make a decision. It could take the form of another controller entirely, such as a conventional PI controller. To avoid this problem the controller should be started with a complete initial model. This model may be inaccurate but it will be constantly improved by the on-line identification.

5 Case studies

A series of case studies was carried out to test the use of fuzzy identification both off-line for MBC design, and on-line for controller adaptation. Each of the case studies is now described briefly. Full details are to be found in[2].

5.1 Silica levels in alumina

This study involved the design of two MBCs for the control of the silica level in the product alumina produced by the Bayer process at ALCOA's Kwinana refinery, near Perth, Western Australia. The design of the controllers was based on the identification of fuzzy process models from process input-output data[2].

An important problem in the production of alumina from bauxite using the Bayer process is silica contamination in the product. The control of silica levels is a very complex problem with many potential solutions. Both the model-based controllers proved to follow closely the control strategy of human operators. However, they were more consistent than the operators, and the response of the controllers was easily changed by altering the shape of the fuzzy performance measure sets. The controllers were run off-line to act as advisory programs to the operators.

5.2 A mass storage tank

The initial tests of the adaptive fuzzy controller were carried out on a computer simulation of a mass storage tank[4]. This was a first-order process with varying gain and time constant. The outlet flow was assumed to be turbulent, and so proportional to the square root of the height of liquid in the tank. The height itself could be manipulated by modifying the inlet flowrate. The aim of the controller was to maintain the height at a specified setpoint.

The MBC used a process model consisting of rules of the form:

If current change in error
And current change in process input
Then next change in error

At each time step, the MBC applied each of a set of seven possible changes in control action (process input) to this model to get predictions for the expected changes in error over the next sample time. These were added to the current error to get predicted values for the error at the next sample time. The change in control action that gave the smallest possible error, of the same sign as the current error, was chosen as the current change in control action. The criterion that the predicted error be the same sign as the current error was introduced to reduce overshoot during setpoint change responses.

An initial model was constructed using logical considerations of a first-order process. The MBC, using this model, was only tuned for specific heights, as the gain of the process changed with the height. It performed best in the upper section of the tank. Response to setpoint changes was somewhat sluggish towards the bottom of the tank, where the gain was lower.

The adaptive controller was trialled on setpoint changes near the top and bottom of the tank. It was started with the initial model, but was able to modify the model on-line to adapt the response to the changing gain of the process. When changing from the top to the bottom of the tank, or vice versa, a single setpoint change was sufficient to adapt the response. The response of the initial

MBC is compared with that of the adaptive controller, for setpoint changes in the bottom of the tank, in Fig .3.

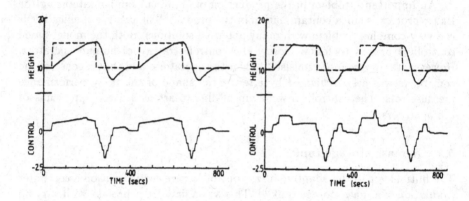

Fig. 3 Setpoint changes near bottom of the tank (a) initial non-adaptive MBC (b) adaptive fuzzy controller.

5.3 A cement kiln

The adaptive fuzzy controller has been applied to the control of a cement kiln simulation. This is a multivariable process requiring a multivariable controller due to the interactions between the variables. The fuzzy adaptive controller was applied to the regulatory control problem in which the measured variables had to be driven to, and maintained at, setpoint.

The cement kiln simulation is a three-input, three-output system, obtained by an empirical fit to plant data. It describes the normal operation of the kiln around the required setpoints. It adequately characterises the interaction between variables that makes the use of individual control loops unsatisfactory. The three measured variables are the burning zone temperature (BZ), the back-end temperature (BE), and the percentage of oxygen in the exit gas (OX). Except under emergency or upset conditions these three variables can be maintained within reasonable limits through adjustment of three manipulated variables, namely the kiln feedrate (KF), the fuel flowrate to the burner (BF), and the speed of the induced draught fan (FS).

The adaptive fuzzy controller was used in a learning mode. It was started without a process model, and had to learn its control strategy by identifying the process model on-line. During the initial learning stages the model-based algorithm could not choose a control action because of the lack of entries in the process model. In this case an auxiliary control algorithm had to be used to generate a control action. The auxiliary algorithm, though less sophisticated than the model-based algorithm, tried to maintain at least crude control of the process until the model-based algorithm could take over. It operated in a similar fashion to the model-based algorithm, but using a model that only contained information about the likely direction of change of variables. This model contained inaccuracies due to assumptions made about the summation of individual changes to the variables. The intent of the algorithm was not only to maintain control of the process, but to stimulate learning in the model-based controller by the use of new control actions.

The fuzzy adaptive controller with the auxiliary algorithm proved to be an effective multivariable learning controller. Control of the cement kiln was maintained, while the learning controller provided a general increase in the quality of control. The final controller response was equivalent to that obtained using a predefined model. The first, third and final (27th) learning runs are shown in Fig .4.

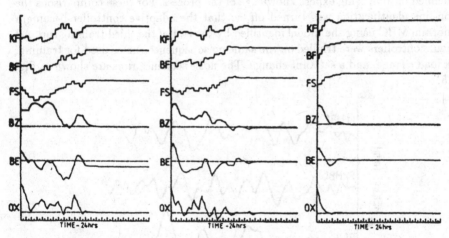

Fig. 4 Setpoint change learning runs for the cement kiln (a) first (b) third (c) final (27th) run.

5.4 The liquid level rig

Experimental trials of the adaptive fuzzy controller were carried out on a laboratory-scale liquid level rig [3]. The liquid level rig consisted of a main water holding tank which was fed with water from the top, and drained of water through a valve at the bottom. Water was circulated continuously by an electric pump. The height of water in the tank was measured by a pressure transducer attached to the bottom of a standpipe connected to the main tank by a small tube. There was a significant time delay involved in the measurement due to the constriction of the tube. The inlet flowrate to the main tank could be adjusted by a valve on the inlet flow side of the pump. This flowrate was measured by a flow meter. The control problem was to control the height of water in the tank by adjusting the outlet flowrate. Variations in the inlet flowrate formed a measurable disturbance. The controller was implemented on an NCR Decision Mate personal computer interfaced to the rig.

The adaptive fuzzy controller was again used in a learning mode, and was started with no model at all. It was trained using a noise sequence on the inlet flowrate that consisted of a normally distributed sequence of random numbers that gave up to a 15 percent change in the inlet valve position. Each training run lasted for five minutes, with a sample time of five seconds. The process was initially at steady-state. The model identified by the end of each run was used as the initial model on the next run. If the MBC could not make a control decision, because the process model contained too few entries, the initial steady-state outlet valve position was used. Training was essentially complete after three runs.

The performance of the adaptive fuzzy controller was then compared with a standard PI controller, a PI / feedforward controller, and a MBC using a model defined off-line using expert knowledge of the process. For these comparisons the on-line identification was turned off, so that the adaptive controller became a normal MBC using the model identified by the end of the third training run. All four controllers were trialled on the same noise sequence run as used for training, a load change, and a setpoint change. The noise sequence runs are shown in Fig. 5.

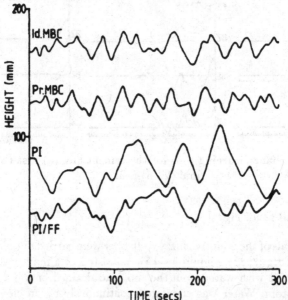

Fig. 5 Comparison of (a) MBC with identified model (b) MBC with predefined model (c) PI and (d) PI/feedforward controllers for a noise sequence run on liquid level rig

In all cases the MBC using the identified model outperformed the PI and PI / feedforward controllers, especially for the noise sequence on which it was trained. It also outperformed the MBC using the predefined model, except for the load change, where the predefined MBC was marginally better on the basis of a smaller height variance.

5.5 A single effect forced circulation evaporator

A novel design technique for the development of a multivariable MBC was explored for the control of a single effect forced circulation evaporator. The evaporator provides a two-input, two-output system that is highly interacting and nonlinear, hence difficult to control. The control scheme consists of controlling the product composition (X2) and the operating pressure (P2) by manipulating the steam pressure (P100) and the cooling water flowrate (F200).

The design technique first develops a crude multivariable fuzzy model by the expansion of single-input, single-output models. The resultant multivariable MBC provides a control response of the same quality as provided by single loop control. Differences in control are due to the multivariable controller choosing changes to the manipulated variables in tandem, rather than independently, as in the single variable controller. By the use of adaptive fuzzy control, the multivariable model is improved, with corresponding refinement of the control response.

The results show that this is a feasible approach to the development of a multivariable MBC. Even in the presence of considerable model mismatch errors, the adaptive fuzzy controller was able to improve the control response of the initial multivariable controller. The single variable, initial multivariable and adapted multivariable responses to a setpoint change in both measured variables is shown in Fig .6.

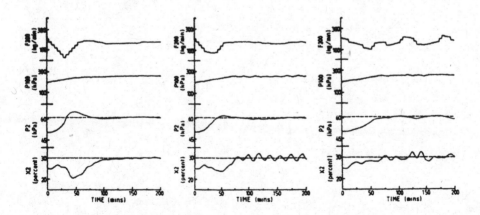

Fig. 6 Setpoint changes for the evaporator (a) two single variable MBCs (b) initial multivariable MBC (c) adapted multivariable MBC.

References

1. Graham, B.P.: Fuzzy identification and control. PhD. Thesis, University of Queensland, (1986).
2. Graham, B.P., Newell,R.B., LePage,G.P., Stonehouse,L.C.: Industrial applications of fuzzy identification and control. Proceedings of 14th Australian Chemical Engineering Conference, pp221-226, (1986).
3. Graham, B.P., Newell,R.B.: Fuzzy identification and control of a liquid level rig. Fuzzy Sets and Systems, 26:255-273, (1988).
4. Graham, B.P., Newell,R.B.: Fuzzy adaptive control of a first-order process. Fuzzy Sets and Systems, 31:47-65, (1989).
5. Hendy, R.J.: Supervisory control using fuzzy set theory. PhD. Thesis, University of Queensland, (1980).
6. Mamdani, E.H.: Application of fuzzy algorithms for control of simple dynamic plant. Proceedings of the Institution of Electrical Engineers, 121:1585-1588, (1974).
7. Zadeh, L.A.: Fuzzy sets", Information and Control, 8:338-353, (1965).

Part III

Fuzzy Logic-based Reasoning and Control: Theoretical Aspects - Fuzzy Neural Networks

Fuzzy Representations in Neural Nets

James Franklin

School of Mathematics, University of New South Wales, P.O. Box 1, Kensington 2033, Autralia

Abstract. Clear, crisp, precise and unambiguous: that is how you like your concepts, if you are a serial computer. But human concepts are in general vague, fuzzy or subject to borderline cases. Anyone who deals with information via computers knows the problems arising from having to categorise objects to fit the computer's crude pigeonholes, and how inflexible this is compared to what humans do. Conversely, those of us who teach mathematics and related subjects know how hard it is to induce the brain to represent clear and precise concepts.

1 Fuzzy Logic

The formalism of fuzzy logic, or fuzzy set theory, aims to represent fuzzy concepts in a form understandable by a computer. The obvious idea on how to tell a computer about a fuzzy concept like "tall" is to give the computer numbers between 0 and 1, which describe the degree to which a borderline case is tall. Someone of height 140 cm is definitely not tall, so has degree of tallness 0, the average professional basketballer has degree of tallness, or "membership in the set of tall people", 1, while someone of height 177 cm is tall to degree, say, 0.4. Since the aim is to imitate natural language, it is envisaged that these numbers will be elicited from speakers of the language by a survey or dialogue. A "membership function" will result, like:

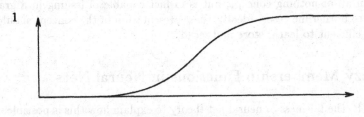

Fig. 1

Given a database containing names of people with their heights in centimetres, the computer should respond to a request like, "List all the tall employees we have in accounts" by something like: "Those more than 0.8 tall are . . . ; do you want those a little less tall as well?". To do this, the computer must have the fuzziness of the concept "tall" explicitly represented, either by a table of values for the membership function, or a formula for it. Then there must be transformations that relate the function for "tall" to those for "fairly tall", "very tall" and so on, and suitable ways of dealing with the conjunction, disjunction and so on of fuzzy concepts (surveys in [17], [42] [37].

This way of dealing with the fuzziness of concepts has not had notable success in rendering interfaces "intelligent", but there seems no obvious alternative within the framework of serial computers. There may eventually be an acceptable outcome, if one is prepared to put in enough work to hand-craft the membership functions to accord with psychological and linguistic reality (which is unexpectedly subtle: [46, 31, 32, 50]. Nevertheless, if one compares this approach with what the brain does, one notices at once certain oddities. The first is a conceptual one. Are the numbers in the fuzzy membership functions themselves crisp or fuzzy? If one says, "Perhaps someone 177cm tall is not 0.4 tall, but only 0.39741 tall", one realises that someone has cheated by taking the examples of fuzzy membership values to one decimal place. Numbers to one decimal place, in ordinary language, *are* fuzzy. It is possible to have a theory of "second-order fuzzy sets" [17], in which the membership values themselves do have a fuzzy membership function, but one seems to be getting further away from reality in doing so. The most offensive aspect of this, from the computing point of view, is that the more vagueness there is, the more is explicitly represented in the computer, by way of explicit membership functions. It has not been easy to implement fuzziness in databases in a way that avoids excessive computation of membership functions[4]. This is the opposite of what happens in the brain (or, for that matter, in decimal numbers) where it is *accuracy* that is expensive.

One is now in a position to ask the question (one unfortunately rarely asked): "What is vagueness or fuzziness *for* ?" The answer is: to save wasting computational resources on pointless accuracy. Properties which in the real world vary continuously, like tallness, could be represented by the brain wastefully by some kind of graduated scale, like a thermometer, which would classify any height-of-person input into the correct division. But it would be better to have a single discrete concept, a prototype or stereotype, "tall", which is to a first approximation an all-or-nothing concept, but is in fact capable of issuing in a graded response to borderline cases. Ideally, the representation of the concept should be easy and efficient to learn, store and recall.

2   Fuzzy Membership Functions in Neural Nets

It should be the business of neural net theory to explain how this is possible, and to explain it well enough to allow realistic implementation of fuzzy concepts.

Here is a very simple example of a neural net, intended to illustrate how a net computes what is in effect a fuzzy membership function, even when it has been trained on discrete data which do not mention fuzziness. It is a net trained to solve the XOR (exclusive or) problem, one of the simplest (and of course very easy) classic problems. It should accept two inputs, which may be 0 and 1 in any of the four possible combinations, and produce a single output: nearly 0 if the inputs are the same, and nearly 1 if the inputs are different.

The network is shown in fig. 2. At the bottom layer are two input neurons (the circles) and one "bias neuron" (the square) which emits a constant 1. The two input neurons are "dummy" neurons: they simply accept their input and pass it on. The top two neurons take the weighted sum of their inputs, transform by the "squashing" function:

$$\sigma(x) = \frac{1}{1 + e^{-x}}$$

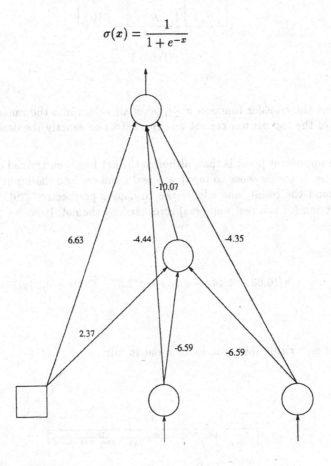

Fig. 2 Net to solve XOR problem: weights after training.

and output the result. The weights shown on the connections have been arrived at by a training algorithm, in this case standard backpropagation. The details of this are not important: they are simply a way of adjusting the parameters of the system, the weights, to achieve the least possible error. The results of the training are:

Input	Desired output	Actual output
0 , 0	0	.07
0 , 1	1	.89
1 , 0	1	.89
1 , 1	0	.10

Table 1

(Since the transfer function σ squashes all values into the range 0 to 1, the output of the top neuron cannot be expected to be exactly the desired values 0 or 1).

The important point is that, although the net has been trained on non-fuzzy 0-1 values, it makes sense to input *any* real numbers into the input neurons. To understand the result, one adopts the "mapping perspective"[29], and asks for the function (of two real numbers) computed by the net. It is:

$$\sigma((6.63 - 4.44x - 4.35y - \sigma(2.37 - 6.59x - 6.59y))$$

That is, writing the formula for σ out in full:

$$\frac{1}{\left[1 + e^{4.44x - 6.63 + 4.35y + \frac{10.07}{1 + e^{6.59x - 2.37 + 6.59y}}}\right]}$$

One understands such a function best by drawing its contours. The 0.8 and 0.2 contours are shown in fig. 3. (Only inputs in the unit square are shown: since σ transforms all large input values to nearly 1, large inputs are not of interest. From the fuzzy point of view too, it is inputs between 0 and 1 which are relevant). The contours are in fact almost straight lines, which is an unusual feature of this example.

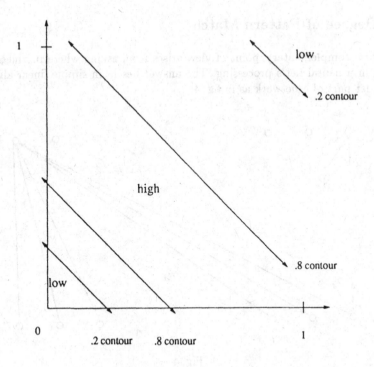

Fig. 3 Output of Network in Fig. 2.

Note especially the smooth transition of the computed output between the high and low values induced by training.

The phenomena shown here, including the smooth transition between 0 and 1 outputs, generalise to more realistic cases, where the relation between input and output is more complex, and the number of neurons used is large.

Neural nets work by "approximation by superpositions of a sigmoidal function" [6]. To classify inputs into two classes, for example, they approximate the indicator (or characteristic) function of one of the classes (that is, the function that is 1 on one class and 0 on the other). Since sigmoidal functions, and therefore their compositions, are continuous and smooth, the transition from one class to the other (across the "decision surface") is smooth. That is, a neural net classifier computes a fuzzy membership function for the decision classes, whether or not the data are given as fuzzy. With longer training and (especially) many more neurons, the steepness of the transition can be increased, but the fuzziness never disappears.

3 Degree of Pattern Match

Another, complementary, point of view arises from asking where fuzziness first arises in a neural net's processing. The answer lies in in simple linear algebra. Consider part of a network as in fig. 4:

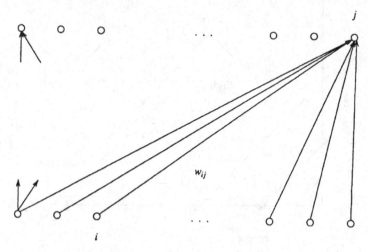

Fig. 4

A pattern of numbers (possibly any real numbers, or possibly only 0's and 1's) is output by the bottom layer of n neurons. The i'th neuron emits the output S_i. The vector of these outputs is $\mathbf{S} = (S_1, \ldots, S_n)$. These are weighted by the weights on the connections: if the weight on the connection between the i'th bottom neuron and the j'th top neuron is w_{ij}, then the signal passing along this connection will be $S_i w_{ij}$. So the total signal received by the j'th top neuron will be $sum_i S_i w_{ij}$. The top neurons (which may be a different layer of neurons, or simply the bottom ones again at a later time) then calculate their outputs according to some rule. Let us concentrate on the expression for the total input into the j'th neuron:

$$\sum_i S_i w_{ij} = \mathbf{S} \cdot \mathbf{w}_j = \parallel \mathbf{S} \parallel \parallel \mathbf{w}_j \parallel \parallel \parallel cos(\mathbf{S}, \mathbf{w}_j)$$

where $\mathbf{w}_j = (w_{1j}, \ldots w_{nj}))$ is the vector of connections into the j'th top neuron[5, 15]. In this equation, the terms $\parallel \mathbf{S} \parallel$ and $\parallel \mathbf{w}_j \parallel$ are just the strengths or sizes of the vectors $\mathbf{S}$ and $\mathbf{w_j}$; the important term is the cos of the "angle" between them. This gives a measure of the match between the two vectors, namely, between the vector of signals from the bottom layer and the vector of weights on the incoming connections to the j'th top neuron.

The j'th neuron computes its output as some monotonic increasing function of the total input, so it is essentially responding to the degree of match between the output pattern from the bottom layer and the pattern of weights on the connections into itself. That is, possible patterns, or "microfeatures", of inputs are represented explicitly in the interconnection patterns. In fact, "represented" is too weak a word: the microfeature is literally in the weights, in the sense that a vector of incoming weights *boldw* may be the *same* as a signal vector **S**. The soul is in a way all things, as Aristotle says (Aristotle -350, bk 3 ch. 4). Individual neurons respond to the particular microfeatures represented in the fan of incoming connections. What they are actually responding to is something that varies effectively continuously (for large n, at least), since $cos(\mathbf{S}, \mathbf{w}_j)$, measuring the degree of pattern match, varies continuously.

We have therefore found how to represent microfeatures in a network, in such a way that the output from them is fuzzy. What implications does this have for the understanding of fuzzy representation of larger features? It is not clear.

If we consider the outputs of larger systems which can be understood in the same way, the Linear Associative Memories (references in [5, 18]), we can find essentially the same phenomenon, of response graded by the degree of pattern match between input and memory.

In these, the situation is as in fig. 4, with the lower row of neurons being an input layer and the upper row an output layer. The output of each top neuron is just the sum of its inputs. (This is what makes these systems linear).

In this simple case, it is easy to say what the weights should be in order to cause the network to output desired vectors $\mathbf{b}_1, \ldots, \mathbf{b_n}$ respectively when given vectors $\mathbf{a}_1, \ldots, \mathbf{a_n}$ as input. If the $\mathbf{a}_i$ are orthogonal, the matrix of weights $(w_i j)$ should be $\sum_k \mathbf{a}_k^t \mathbf{b_k}$ (the vectors being regarded as row vectors). Inputting $\mathbf{a}_i$ into the system with these weights then causes (a multiple of) the corresponding $\mathbf{b}_i$ to be output. Further, inputting a vector close to $\mathbf{a}_i$ causes a vector close to $\mathbf{b}_i$ to be output. The reason is that the case here is simply a heap of the cases with one neuron considered above: each top neuron calculates its own degree of pattern match, and outputs the result without processing. Thus the network does incorporate a fuzzy response to output.

Naturally, things are less clear in the more realistic cases where the responses of neurons are non-linear functions of their input, and where there are feedbacks. But one can at least say that in a simple feedforward network of the sort in the example above, the sigmoid or squashing function preserves the graded response of the neuron to its microfeature: a squashed graded response is still graded (unless the input is very large). It is possible to study how much fuzziness remains in the system by following the propagation of errors in input through the nodes.

What, then, of fuzzy logic? The aim is to produce a system in which concepts or features of the world are represented in such a way that the output from stimulation of a memory of the concept is fuzzy. That is, it varies according as the new input matches the input which originally gave rise to the memory. The fuzziness should not be represented explicitly inside the system, as the fuzzy logic project originally assumed, but should issue from the way inputs interact with stored patterns.

4 Cluster Analysis and Fuzzy Control Rules

Some comments are in order on the subject of cluster analysis, since it is often said that neural nets are an architecture for clustering data[22, 29, 16]. It has also been observed that simple unsupervised nets automatically perform principal component analysis, a simple clustering technique which finds structure in high-dimensional data sets [28, 23, 2, 35]. Training algorithms for neural nets, like clustering algorithms, use input data to adjust the internal parameters of a classification system. The trained system is then intended to classify both old and new data correctly. The clustering problem is often presented as that of tessellating or "vector quantising" the input space into decision classes, in accordance with the distribution of the input data. If seen this way, there would be no fuzziness involved. But this misrepresents what most clustering algorithms actually do. For example, the classical K-means clustering algorithm[24, 19] divides the space into K classes according to the distance of points from K prototype vectors; the positions of these K vectors are decided by the input data. A division of the space like that in fig. 5 (for $K = 3$) results:

5 Decision Regions with three Prototype Vectors

One loses important information by keeping only the tessellation, ignoring the continuous functions on which it is based (the distances of points from the three prototype vectors). Points near the decision boundaries should be regarded as borderline cases, since there the distances to two (or three) prototypes are nearly equal. Thus a scoring rule (for deciding how well the partition is performing) should not penalise mistakes near the boundaries as heavily as ones away from them. And in a control situation, where the classification is used to decide on action, one should not have a sudden change in the action as the decision boundary is crossed.

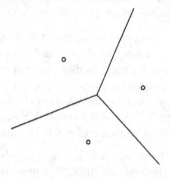

Fig. 5

This suggests that it would be best to control by fuzzy control rules (in the Japanese sense), which are adapted to the way the input data cluster. In the centre of a cluster, one rule should apply, but near the boundary between clusters, the outputs of the different rules should be suitably averaged, so that a smooth transition between rules results. (This is the basic idea of fuzzy control: [21, 48]. This has been largely achieved in the Fuzzy Associative Memory of Bart Kosko[20]. Some similar work has been done also by Takagi and Hayashi[43], who point out that using neural nets to adapt control rules solves two long-standing problems for conventional fuzzy reasoning: the "lack of design for a membership function except a heuristic approach" and "the lack of adaptability for possible changes in the environment". It is possible to hope that the same approach may also create the philosophers' stone of neural net theory, the transition from a black-box input-output system to a symbol manipulator. A cluster is a discrete entity, like a symbol, and a fuzzy rule links two such clusters, in a way similar to the logical if-then relation between symbols.

Neural nets for clustering share a problem long recognised as a difficulty for clustering algorithms generally - that of deciding how many clusters (or neurons) is appropriate [8].

6   The Psychological Evidence

There are various reasons why one should study how the brain performs on categorisation tasks, and how well its performance may be simulated by neural net models. One is the hope of understanding human cognition. While much can be expected in the long term, at present neural net models seem too primitive to give a real level of understanding - if a net of ten neurons produces some behaviour which appears to resemble some aspect of what the brain does with billions of neurons, it is hard to make a case that real understanding has been achieved. The point of view of this paper is different. It is that fuzzy logic aims to imitate human concepts as simply as possible, and that neural nets provide a way of doing so, which is both theoretically and practically easier than hand-crafting membership functions. One should therefore examine human performance, in order to engineer neural networks to perform similarly. Work in psychology has focussed on supervised learning of classification. Human subjects are trained on examples, being told which of a variety of examples presented belong to different classes. The subjects are then asked to classify new examples, which differ somewhat from all the training examples. Psychologists considered two extreme models, the prototype and the exemplar models. According to the first, subjects would store a single prototype of each class, a kind of average of the training examples for that class; they would classify new cases solely on the basis of distance from the nearest prototype (much in the fashion of the K-means clustering discussed above). On the exemplar model, subjects would store all the training examples, and classify new cases on the basis of the distance to the nearest exemplar. Not surprisingly, it was discovered that neither model fitted the evidence very well, and results fell somewhere between the two (though the prototype model worked quite well for younger children) (review in [41]. This is what would be expected from a neural net model, which neither keeps all the examples, nor collapses the concept to a single point (classifying on the basis of distance to a single prototype means that the contours of the membership function of the classes are spheres; neural nets do not prefer spheres). One would like to be able to distinguish experimentally between a neural net model and a mixture of the prototype and exemplar models. This seems in principle difficult, if not impossible, since the neural net model *is* a kind of mixture of the prototype and exemplar models. Nevertheless, there has been some relevant work done ([7, 39]).

Some other fuzzy phenomena that one would want to replicate with a neural net categorisation model include the variation among exemplars of a category as to their "prototypicality" [34] and the "exponential-decay relationship between the probability of generalization and psychological distance" [10].

One important part of the psychological evidence is the finding that prototypes are very subject to the effects of background knowledge and context[12, 27]. For example, the bird one thinks of in "The bird sang" is different to the bird one thinks of in "The bird walked across the barnyard". In principle, a reasonably large neural net should be able to handle this, since neural net learning is always in a context (of prior weights), and thus performs a kind of quasi-Bayesian modification of categories, adjusting old categories in the light of new examples[30]. But it cannot be said that this can be engineered at present.

7 Other Issues: Noise and Distributedness

The above fails to deal with certain issues which seem intuitively essential to understanding why neural nets issue naturally in a fuzzy response. They are:

- The performance of neural nets is said to "degrade gracefully in the presence of noise". Noise is something unintended, while borderline cases are envisaged as normal, but it would be expected that the mechanisms of continuous variation in response might be the same in each case. Graceful degradation, while much commented on, is not well understood mathematically (that is, is not understood at all); the little work there has been has mostly centred on the effects of noise in training data: [11, 47].
- A memory's being distributed seems intuitively to allow it to have a graded response, for example to damage, in a way that a localised memory, as in a conventional single site in a memory accessed by address, cannot possibly have. A memory on disk is either there or not, whereas a distributed memory can be partly there.
-. The somewhat obscure connection between distributed memories and holographs[33] promised to at least provide a mathematically comprehensible model of how output can degrade partially: a holograph with a small piece removed produces an image which is fuzzy, but whole.

These phenomena make it clear, too, why fuzziness is a topic central to understanding neural nets, whereas in the serial case it is peripheral, and a nuisance.

Understanding distributed representation in neural nets is difficult because representation of concepts is quite different in a neural net to a serial computer memory. If one asks, Why is this sequence of 0s and 1s on this disk a representation of that bit of the world?, one answers, Because the programmer said so. Reference is by stipulation, or convention. (There are some exceptions, in that structural aspects can actually be in the computer: for example, in a computer simulation of city growth, there may be sequences of 0s and 1s representing houses, and the structure of the ensemble of these should be the same as that of the city: that is what simulation means. Nevertheless the link between the atomic sequences and the individual houses is purely conventional). This makes representation in the serial case easy, in one way, in that it is reduced to representation in the human programmer, but harder in another way, in that the human case is a problem itself, as any book on the philosophy of language reveals.

The matter is different in the neural net case. A neural net usually acquires structure by training, not programming, and the representation is somewhere in this structure. Therefore the relation of stipulation is replaced by one of causality. The structure of the part of the world being represented is directly causing the structure that the neural net is acquiring. Philosophically, this is much simpler, being just the same as changes in temperature causing the mercury to vary concomitantly in a thermometer. In practice, of course, it is much harder. All one can see is a vast number of interconnection weights changing in response to inputs, and one loses track completely of what patterns in the weights have been caused by what patterns in the inputs[14, 36, 38, 13, 45].

What is notably missing from the approach so far is an explanation of how the mere fact of a representation's being distributed leads to fuzzy output. The simplest way to create a distributed representation of an object is to make a number of copies of whatever represents it. The fuzziness of response to the concept would then reside in the proportion of these modules activated. Thus if a number of neurons represented the concept "tall", a borderline case of tallness would activate only some of them. This feature is preserved when, in the interests of efficiency, these neurons are used in the representation of various other concepts as well (as in [44]). This is what happens with the genetic code. The relationship between sections of the DNA and characteristics of the organism is nothing like one-to-one: the coat colour of mice, for example, is affected by fifty different parts of the DNA, most of which have other gross effects[40]. Intuitively, the many-to-many relationship would be expected to lead to both resilience to damage and to continuous variation in expressed characteristics, but the matter cannot be called well understood.

References

1. Aristotle: On the Soul. (350 BC).
2. Baldi, P. Hornik, K.: Neural networks and principal component analysis. Neural Networks 2: (1989) 53-58.
3. Bezdek, J.C.: Pattern Recognition with Fuzzy Objective Function Algorithms Plenum, New York, (1981).

4. Bosc, P., Galibourg, M., Hamon, G.: Fuzzy querying with SQL: extensions and implementation aspects. Fuzzy Sets and Systems 28: (1988) 333-349.
5. Carpenter, G.A: Neural network models for pattern recognition and associative memory. Neural Networks 2: (1989) 243-257.
6. Cybenko, G.: Approximation by superpositions of a sigmoidal function. Mathematics of Control, Signals and Systems 2: (1989) 303-314.
7. Estes, W.K., Campbell, J.A., Hatsopoulos, N., Hurwitz, J.B.: Base-rate effects in category learning: a composition of parallel network and memory storage-retrieval models. Journal of Experimental Psychology: Learning, Memory & Cognition 15: (1989) 556-571.
8. Fritzke, B.: Unsupervised clustering with growing cell structures. Proceedings of the IJCNN-91, Seattle, (1991).
9. Funahashi, K.-I.: On the approximate realization of continuous mappings by neural networks.0 Neural Networks 2: (1989) 183-192.
10. Gluck, M.A.: Stimulus generalization and representation in adaptive network models of category learning. Psychological Science 2: (1991) 50-55.
11. Györgyi, G.: Inference of a rule by a neural network with thermal noise. Physical Review Letters 64: (1990) 2957-2960.
12. Hayes, B.K., Taplin, J.E.: The effects of existing knowledge on the learning of artificial categories. in: (A.F. Bennett & K.M. McConkey, eds), Cognition in Individual and Social Contexts North-Holland, (1989).
13. Herz, A., Sulzer, B., Kühn R., van Hemmen, J.L.: Hebbian learning reconsidered: representation of static and dynamic objects in associative neural nets. Biological Cybernetics 60: (1989) 457-467.
14. Hinton, G.E., McClelland, J.L., Rumelhart, D.E.: Distributed representations. Chapter 3 of D.E. Rumelhart & J.L. McClelland, Parallel Distributed Processing, MIT Press, 1986.
15. Jordan, M.I.: An introduction to linear algebra in parallel distributed processing. Chapter 9 of D.E. Rumelhart & J.L. McClelland, Parallel Distributed Processing, MIT Press, 1986.
16. Kamgar-Parsi, B., Gualtieri, J.A., Devaney, J.E., Kamgar-Parsi, B.: Clustering with neural networks. Biological Cybernetics 63: (1990) 201-208.
17. Klir, G., Folger, T.: Fuzzy Sets, Uncertainty and Information, Prentice-Hall, (1988).
18. Kohonen, T.: Content Addressable Memories. Springer Verlag, (1980).
19. Kohonen, T.: Self-Organization and Associative Memory 3rd edition, Springer Verlag, (1989).
20. Kosko, B.: Neural Networks and Fuzzy Systems. Prentice Hall, 1991.
21. Lee, C.C. Fuzzy logic in control systems', IEEE Trans. on Systems, Man and Cybernetics 20: 404-418 (part 1), 419-435 (part 2) (1990).
22. Lippmann, R.P.: An introduction to computing with neural nets. IEEE ASSP Magazine 4(4): 4-22, (1987).
23. Linsker, R. Self-organization in a perceptual network. Computer 21(3): (1988) 105-117.
24. MacQueen, J.: Some methods for classification and analysis of multivariate observations. Proceedings of the Fifth Berkeley Symposium on Mathematical Statistics and Probability vol I pp. 281-297, (1967).
25. McClelland, J.L., Rumelhart, D.E. Distributed memory and the representation of general and specific information. Journal of Experimental Psychology: General 114: (1985) 159-188.

26. Mira, J., Delgado, A.E., Moreno-Diz: The fuzzy paradigm for knowledge representation in cerebral dynamics. Fuzzy Sets and Systems 23: (1987) 315-330.

27. Murphy, G.L., Medin, D.L.: The role of theories in conceptual coherence. Psychological Review 92: (1985) 289-316.

28. Oja, E.: A simplified neuron model as a principal component analyzer. Journal of Mathematical Biology 15: (1982) 267-273.

29. Pao, Y.-H.: Adaptive Pattern Recognition and Neural Networks Addison-Wesley, (1989).

30. Poggio, T., Girosi, F.: Regularization algorithms for learning that are equivalent to multilayer networks. Science 247: 978-982 (1990).

31. Pohl, N.F.: Scale considerations in using vague quantifiers. Journal of Experimental Education 49: (1980) 235-240.

32. Powell, M.J.: Purposive vagueness. Journal of Linguistics 21: (1985) 31-50.

33. Pribram, K.H.: Languages of the Brain. Prentice Hall, (1971).

34. Rosch, E., Mervis, C.B.: Family resemblances: studies in the internal structure of categories. Cognitive Psychology 7: (1975) 573-605.

35. Rubner, J., Taran, P.: A self-organizing network for principal-component analysis. Europhysics Letters 10: (1989) 693-698.

36. Rumelhart, D.E., Hinton, G.E., Williams, R.J.: Learning internal representations by back-propagating errors. Nature 323: (1986), 533-536.

37. Rundensteiner, E.A., Hawkes, L.W., Bandler, W.: On nearness measures in fuzzy relational data models. International Journal of Approximate Reasoning 3: (1989) 267-298.

38. Saund, E.: Abstraction and representation of continuous variables in connectionist networks. in: Proceedings AAAI-86, Philadelphia: (1986) 638-644.

39. Shanks, D.R.: Connectionism and the learning of probabilistic concepts. Quarterly Journal of Experimental Psychology 42A: (1990) 209-237.

40. Silvers, W.K.: The Coat Colours of Mice: A Model for Mammalian Gene Action and Interaction, Springer Verlag, (1979).

41. Smith, E.R., Zarate, M.A.: Examplar and prototype use in social categorization. Social Cognition 8: (1990) 243-262.

42. Smithson, M.: Fuzzy Set Analysis for Behavioural and Social Sciences, Springer Verlag, (1987).

43. Takagi, H., Hayashi, I.: NN-driven fuzzy reasoning. International Jopurnal of Approximate Reasoning 5: (1991) 191-212.

44. Touretzky, D.S., Hinton, G.E.: Symbols among the neurons. Proceedings of 9th International Joint Conference on Artificial Intelligence Vol. 1: (1985) 238-243.

45. Touretzky, D.S., Pomerleau, D.A.: What's hidden in the hidden layers. Byte (Aug): (1989) 227-233.

46. Wierzbicka, A.: Precision in vagueness: the semantics of English approximatives. Journal of Pragmatics 10: (1986) 597-614.

47. Wong, K.Y.M., Sherrington, D.: Training noise adaptation in attractor neural networks. Journal of Physics and Applied Math. Gen. 23: (1990) L175-L182.

48. Yamakawa, T.: Stabilization of an inverted pendulum by a high-speed fuzzy logic controller hardware system. Fuzzy Sets and Systems 32: (1989) 161-180.

49. Young. S.W., Ballerio, C., Carrol, C.L.: Visual fuzzy cluster-analysis of MR images. American Journal of Roentgenology 152: (1989) 19-25.

50. Zimmermann, H.-J., Zysno, P.: Quantifying vagueness in decision models. European Journal of Operational Research 22: (1985) 148-158.

A Method to Implement Qualitative Knowledge in Multi-Layered Neural Network

Hiroshi Narazaki[1],[2] and Anca L. Ralescu[3],[4]

[1] On leave from Electronics Research Laboratory, Kobe Steel, Ltd., Japan
[2] LIFE Chair of Fuzzy Theory, Department of Systems Science, Tokyo Institute of Technology, 4259 Nagatsuta, Midori-ku, Yokohama, 227 Japan
[3] On leave from Computer Science Department, University of Cincinnati, USA.
[4] Laboratory for International Fuzzy Engineering Research (LIFE), 89-1, Yamashita-cho, Naka-ku, Yokohama, 231 Japan

Abstract. We propose an improved synthesis method for the multi-layered neural network(NN) using a "translation mechanism" that maps the logical representation of qualitative knowledge into a multi-layered NN structure. We give realizability conditions and synthesis equations to realize logical functions by the NN.

The NN is tuned by back-propagation(BP) after the direct synthesis. This direct synthesis decreases the burden on the BP and contributes to the efficiency and accuracy of BP learning process.

We demonstrate our method through function approximation and character recognition problems.

1 Introduction

In this paper we describe a method to translate our qualitative knowledge into the multi-layered NN structure.

The multi-layered neural network(NN) has been successfully applied to many problems such as pattern recognition and control problems[1]-[5].

The major advantages of using the NN in various applications are its interpolation ability and the adaptability by the back-propagation(BP).

The interpolation in the logical and inference domain has been extensively discussed in fuzzy logic. Therefore, it is natural that many researchers work towards the integration of the NN and fuzzy logic, using the NN as a "hardware" for the fuzzy logic[6]-[10]. The NN is expected to represent the quantitative aspect of our qualitative knowledge described in fuzzy logic.

In this paper we discuss the relationship between these two methods. More specifically, we discuss how our qualitative knowledge can be translated into the NN structure, giving some explicit realizability conditions and the NN synthesis equations. In other words, we present a network builder that generates an initial NN based on our qualitative knowledge described by a logical expression. The initial NN is fine tuned by BP to let the NN learn how the interpolation should be done. The direct synthesis contributes to the efficiency and stability of the learning by BP. (The result of NN becomes more predictable, less depending on coincidental factors such as the initial parameter values).

In this paper we first show analytical results. Then we show two examples: (1) a function approximation, especially a membership function estimation, and (2) character recognition.

2 Analytical Results

2.1 Sigmoid Function

The output of a neuron is given by

$$z = f(\sum_{i=1}^{m} w_i x_i - \theta) \tag{1}$$

where w_i is a weight for an input x_i, θ is a constant "threshold", and $f(\cdot)$ is a sigmoid function given by $f(x) = 1/(1 + e^{-kx})$. We suppose that $w_i > 0$ for all i and $\theta > 0$ throughout this paper. We define two parameters for a sigmoid function.

$$\Xi_L^{\epsilon 1} = Sup\{x; f(x) < \epsilon 1\}, \quad \Xi_U^{\epsilon 2} = Inf\{x; f(x) > 1 - \epsilon 2\} \tag{2.1}$$

where $\epsilon 1$ and $\epsilon 2$ are positive numbers describing negligible errors. We say $f(x) = 1$ when $f(x) > 1 - \epsilon 2$ and $f(x) = 0$ if $f(x) < \epsilon 1$. (In multi-layered NN, we have to be aware of the error accumulation as noted later).

By solving (2.1) for x, we have

$$\Xi_L^{\epsilon 1} = -\frac{1}{k} ln(\frac{1}{\epsilon 1} - 1), \quad \Xi_U^{\epsilon 2} = -\frac{1}{k} ln(\frac{\epsilon 2}{1 - \epsilon 2}) \tag{2.2}$$

Hereafter, we assume that $\epsilon = \epsilon 1 = \epsilon 2 < 0.5$ unless mentioned as otherwise. Notice that, under this assumption, we have

$$\Xi_L^{\epsilon} + \Xi_U^{\epsilon} = 0 \tag{3}$$

2.2 Logical Operations with Continuous Input Variables

Here we give the conditions and synthesis equations for logical functions with *continuous inputs* for later use.

We assume $x_i \in [0, 1]$ and associate two parameters, x_i^{Min} and x_i^{Max}, to each x_i. If $x_i < x_i^{Min}$, then we regard x_i as "False" and if $x_i > x_i^{Max}$, "True". We assume that $0 \leq x_i^{Min} < x_i^{Max} \leq 1$ unless otherwise mentioned.

2.3 AND-neuron

The AND-neuron realizes a conjunction $\wedge_{i=1}^{m} x_i$. It is extended so as to output 1 if $x_i \in [x_i^{Max}, 1]$ for all i, and 0 if $x_i \in [0, x_i^{Min}]$ for at least one i. ("Don't care", in between).

Assume that x_i is indexed so that $w_i(1 - x_i^{Min})$ is non-increasing without loss of generality. We define an AND-neuron by the following two inequalities.

$$\sum_{i=1}^{m} w_i x_i^{Max} - \theta > \Xi_U^\epsilon \tag{4.1}$$

$$Max_j\{\sum_{i \neq j} w_i + w_j x_j^{Min}\} - \theta = \sum_{i=1}^{m} w_i - w_m(1 - x_m^{Min}) - \theta < \Xi_L^\epsilon \tag{4.2}$$

By combining (4.1) and (4.2), we have

$$\sum_{i=1}^{m} w_i x_i^{Max} - \Xi_U^\epsilon > \theta > \sum_{i=1}^{m} w_i - w_m(1 - x_m^{Min}) - \Xi_L^\epsilon \tag{5}$$

The above inequality is valid if and only if

$$\sum_{i=1}^{m} w_i x_i^{Max} - \Xi_U^\epsilon > \sum_{i=1}^{m} w_i - w_m(1 - x_m^{Min}) - \Xi_L^\epsilon \tag{6.1}$$

Or, equivalently,

$$w_m(1 - x_m^{Min}) - \sum_{i=1}^{m} w_i(1 - x_i^{Max}) > \Xi_U^\epsilon - \Xi_L^\epsilon \tag{6.2}$$

The following lemma gives the condition for (6.2). (See Appendix 1 for proof.)

Lemma 1(Realizability of AND-neuron)

The inequality (6.2) is satisfiable if and only if

$$\sum_{i=1}^{m} \frac{1 - x_i^{Max}}{1 - x_i^{Min}} < 1 \tag{8}$$

By choosing θ as a midpoint of its permissible interval (5), we have the synthesis equations in Table 1.

	AND-neuron	OR-neuron	M-neuron
Condition	$\sum_{j=1}^{m} k_j < 1$ where $k_j = \dfrac{1 - x_j^{Max}}{1 - x_j^{Min}}$	$\sum_{j=1}^{m} h_j < 1$ where $h_j = \dfrac{x_j^{Min}}{x_j^{Max}}$	$x_i^{Min} < x_i^{Max}$ $(i = 1,2,...,m)$
Weights $\gamma > 1$	$w_i = \dfrac{a}{1 - x_i^{Min}}$ where $a = \dfrac{\gamma(\Xi_U^\epsilon - \Xi_L^\epsilon)}{1 - \sum_{j=1}^{m} k_j}$	$w_i = \dfrac{b}{x_i^{Max}}$ where $b = \dfrac{\gamma(\Xi_U^\epsilon - \Xi_L^\epsilon)}{1 - \sum_{j=1}^{m} h_j}$	$w_i = \dfrac{\gamma(\Xi_U^\epsilon - \Xi_L^\epsilon)}{m(x_i^{Max} - x_i^{Min})}$
Threshold $\gamma > 1$	$\theta = \dfrac{a}{2}\left(\sum_{j=1}^{m} \dfrac{1 + x_j^{Max}}{1 - x_j^{Min}} - 1\right)$	$\theta = \dfrac{\gamma(\Xi_U^\epsilon - \Xi_L^\epsilon)}{2} \times \dfrac{1 + \sum_{j=1}^{m} h_j}{1 - \sum_{j=1}^{m} h_j}$	$\theta = \dfrac{\gamma(\Xi_U^\epsilon - \Xi_L^\epsilon)}{2} \times \dfrac{1}{m}\sum_{i=1}^{m} \dfrac{x_i^{Max} + x_i^{Min}}{x_i^{Max} - x_i^{Min}}$

Table 1 Synthesis equations.

By a "Boolean AND-neuron", we mean the case where $x_i^{Min} = 0$ and $x_i^{Max} = 1$ for all i. In this case, (7) always holds, and thus always realizable.

OR-neuron The OR-neuron is defined to output 1 if at least one x_i exceeds x_i^{Max} and 0 if all inputs are less than x_i^{Min}, i.e.,

$$Min_i\{w_i x_i^{Max}\} - \theta > \Xi_U^\epsilon \tag{8.1}$$

$$\sum_{i=1}^{m} w_i x_i^{Min} - \theta < \Xi_L^\epsilon \tag{8.2}$$

By equivalently combining (8.1) and (8.2), we have

$$Min_i\{w_i x_i^{Max}\} - \Xi_U^\epsilon > \theta > \sum_{i=1}^{m} w_i x_i^{Min} - \Xi_L^\epsilon \tag{9}$$

The inequality (9) is valid if and only if

$$Min_i\{w_i x_i^{Max}\} - \Xi_U^\epsilon > \sum_{i=1}^{m} w_i x_i^{Min} - \Xi_L^\epsilon \tag{10}$$

The following lemma gives the condition for the validness of (10). See Appendix 2 for proof.

Lemma 2(Realizability of OR-neuron)

The inequality (10) is satisfiable if and only if

$$\sum_{i=1}^{m} \frac{x_i^{Min}}{x_i^{Max}} < 1 \tag{11}$$

Similar to the previous AND-neuron, by choosing θ as a midpoint of the interval (9), we obtain the synthesis equations in Table 1.

Again a "Boolean OR-neuron" is defined by $x_i^{Min} = 0$ and $x_i^{Max} = 1$ for all i. In this case, (11) always holds and thus the OR-neuron is always realizable.

M-neuron and I-neuron M-neuron is a mixture of AND and OR functions that outputs 1 when $x_i > x_i^{Max}$, $\forall i$ and 0 when $x_i < x_i^{Min}$, $\forall i$. We call this type of neuron M-neuron (Monotonic-neuron). The formal definitions are

$$\sum_{i=1}^{m} w_i x_i^{Max} - \theta > \Xi_U^\epsilon \tag{12.1}$$

$$\sum_{i=1}^{m} w_i x_i^{Min} - \theta < \Xi_L^\epsilon \tag{12.2}$$

Similar to the above discussions, by combining them, we have

$$\sum_{i=1}^{m} w_i x_i^{Max} - \Xi_U^\epsilon > \theta > \sum_{i=1}^{m} w_i x_i^{Min} - \Xi_L^\epsilon \tag{13.1}$$

This is valid if and only if

$$\sum_{i=1}^{m} w_i(x_i^{Max} - x_i^{Min}) > \Xi_U^\epsilon - \Xi_L^\epsilon > 0 \tag{13.2}$$

Clearly, as shown in Table 1, M-neuron is always realizable as long as $x_i^{Min} < x_i^{Max}$. (There is even no need for limiting their values in [0,1]).

Here also θ is chosen as the midpoint of the permissible range as before.

When we have only one input variable, it gives an increasing curve(I-curve) reaching 1 when $x > x^{Max}$. We call a M-neuron with only one input variable an I-neuron. (Note that I-neuron is equivalent to AND and OR-neurons with only one input variable).

NOT-operation and Consistency Condition If we express $\neg x_m$ by $1 - x_m$, then the input to a neuron is given by

$$\sum_{i=1}^{m-1} w_i x_i + w_m(1 - x_m) - \theta = \sum_{i=1}^{m-1} w_i x_i - w_m x_m - (\theta - w_m) \tag{14}$$

Therefore we can express the negation $\neg x_i$ by the following two transformations.

$$w_i \to -w_i, \quad \theta \to \theta - w_i \tag{15}$$

Notice that the above "NOT-operation" is a rotation around $x = 0.5$. Therefore the new x_i^{Min} and x_i^{Max} after the NOT-operation is given by $1 - x_i^{Max}$ and $1 - x_i^{Min}$. The usual or intuitive interpretation of $\neg x$ is a rotation around $z = 0.5$. As shown in Fig. 1, these two rotations yield different results unless the following consistency condition holds.

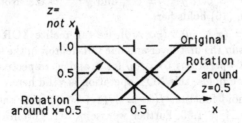

Fig. 1 NOT-operation.

Consistency Condition

$$x_i^{Min} + x_i^{Max} = 1 \tag{16}$$

Note that the consistency condition (16) makes the realizability conditions of AND-neuron in (7) and OR-neuron in (11) identical.

Remarks

1. Any logical formula in a disjunctive normal form (i.e., a disjunction of terms in which each term is a finite conjunction) can be realized by a two layer NN.[5]

   The first layer contains Boolean AND-neurons with appropriate NOT-operations and the second layer contains an OR-neuron.

2. In the case of continuous input variables, we need an additional layer below the AND-neuron layer that contains I-neurons. The I-neuron maps x_i^{Max} and x_i^{Min} to 1 and 0 respectively, and the problem is reduced to the realization of Boolean functions.

3. Care is needed about the accumulation of errors. The error can be more than ϵ for upper layers unless we choose x_i^{Max} and x_i^{Min} to $1-\epsilon$ and ϵ respectively. Note that the realizability conditions require $\epsilon < 1/(m+1)$ for such a choice to be valid.

---

[5] Our first layer contains a sigmoid function, different from the usual input layer" that contains no sigmoid function. However, in the discussions here, we do not count the conventional input layer.

3 Examples

3.1 Exclusive OR

Here we construct the exclusive OR, $x \; XOR \; y$, with the following specifications: $x^{Min} = 0.3$, $x^{Max} = 0.7$, $y^{Min} = 0.2$, and $y^{Max} = 0.8$. Notice that the consistency condition in (16) holds here.

Because $x \; XOR \; y = (x \wedge \neg y) \vee (\neg x \wedge y)$, we can realize XOR function by using two AND-neurons in the first layer and one OR-neuron in the second layer. We have to apply NOT-operations to $x \wedge y$ for x and y, respectively. Because the consistency condition holds, this NOT operation is valid here.

Here we use a sigmoid function $f(x) = 1/(1 + e^{-3x})$ with the error level $\epsilon = 0.01$; thus $\Xi_U^\epsilon = -\Xi_L^\epsilon = 1.53$. Further we use $\gamma = 1$. ¿From the synthesis equations in Table 1, we can determine the parameters as shown in Fig. 2. Notice that the final error is larger than $\epsilon = 0.01$ because of the error accumulation. One compensation method is to use a larger γ for the OR-neuron. The results are shown in Table 2.

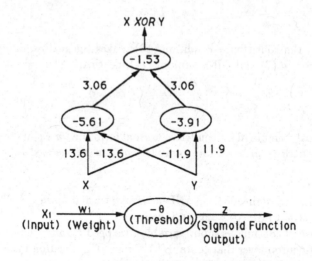

Fig. 2 x XOR y.

X	Y	$\gamma = 1$	$\gamma = 1.5$
0.3	0.2	0.0101	0.00102
0.3	0.8	0.989	0.999
0.7	0.2	0.989	0.999
0.7	0.8	0.0101	0.00102

Table 2 Exclusive OR.

3.2 A Membership Function Estimation and A Function Approximation

The realization of a bell-shaped membership function in Fig. 3 by NN is a direct application of our method. We assume that the universe of discourse is $[0, 1]$ and we are given the following prior knowledge: (1) the center of the fuzzy set in question, i.e., the fuzzy set is described as "Around x= x0", and (2) an initial guess for the support set width l.

Our NN for the above bell-shaped fuzzy set has the structure such that two I-neurons, I_1 and I_2, in the first layer, and one AND-neuron P in the second layer and one NOT-operation to the input from I_2, i.e., $P = I_1 \wedge \neg I_2$. See Fig. 3 for illustration. (x^{Min}, x^{Max}) of the I-curves are $(x0 - l/2, x0)$ and $(x0, x0 + l/2)$.

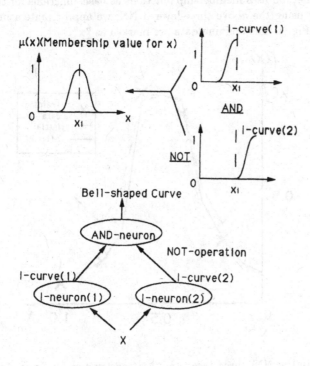

Fig. 3 Fuzzy set realization by NN.

The above initial membership function can be tuned by BP, but, because the interpolation curve is limited to an exponential curve, in many cases BP cannot achieve sufficient fitting accuracy for the training data with the above *minimal* structure. The membership function estimation is a particular case of an approximation problem of a function $y = g(x)$ where $g(x)$ gives a bell-shaped curve. In general, we can use the following approximation.

1. Construct fuzzy sets for *Around* $x = x_i$, $i=1,2,...,k$ (we assume $x_{i+1} > x_i$) using the first and second layers as stated above. (k AND-neurons $\{P_i, i =$

1, 2, ..., k} are arrayed in the second layer). The number and locations of those fuzzy sets, k, determine the accuracy of the approximation. In a single variable case, the support width can be determined as $x_{i+1} - x_{i-1}$ which means that $x = x_i$ belongs only to the fuzzy set *Around* x_i with membership degree 1.

2. Prepare an output neuron Z in the third layer and connect the outputs of AND-neurons {P_i}, i.e., the membership degree of x to fuzzy sets "*Around* x_i" to Z. Determine the weights from P_i to Z, denoted w_i, as $w_i = f^{-1}(g(x_i))$ where $f^{-1}(\cdot)$ is an inverse of a sigmoid function. This weight guarantees that the neuron Z outputs $g(x_i)$ at $x = x_i$. (The value can be deviated due to the error accumulation).

The above method uses membership functions as basis functions for the approximation. By using the above three-layered NN, we approximate a membership function in Fig. 4. The training data are marked as "x".

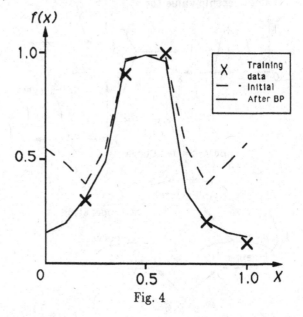

Fig. 4

For the initial NN, three fuzzy sets are constructed for "x is *Around* 0.25, 0.5, and 0.75" and the intended outputs are 0.25, 1.0, and 0.25 at respective points. The initial NN is knowingly deviated from the training data (marked as 'X' in Fig.4) to see how much BP can take care of this mismatch. The initial curve is shown in a dotted line in Fig.4. The increase at the boundary points are due to the zero input to Z. (The boundary points tend to belong to none of the three fuzzy sets). The deviations at $x = 0.25, 0.5, 0.75$ are due to the error accumulation of ϵ. These error accumulations are not important because BP can effectively deal with it. On the contrary, the choice of too small an ϵ is often undesirable for the smooth interpolation and convergence speed because the curve tends to overfit to the training data, especially when the training data are sparse like this example. Therefore this "blurred" initial synthesis is rather desirable. In this example we chose $\epsilon = 0.1$.

After 200 iterations for each training data, i.e., 1000 steps in total, the maximum error was 0.062. The obtained result is also shown in Fig. 4 in a filled line connecting the points at intervals 0.1. This shows the contribution of direct synthesis to the efficiency of the learning process.

3.3 Pattern Recognition

The direct synthesis method will be especially useful for a large NN. We can spare considerable amount of time for BP by starting from a *good* initial NN that at least satisfies nominal specifications. Here we show another example. The purpose is to construct a NN that recognizes three characters, u, v, and w. The two dimensional input plane $X \times Y$ is divided into a grid by dividing each axis into 5 segments. Thus we have $5 \times 5 = 25$ cells, written as $\{n_{ij} = (x_i, y_j), i, j = 1, 2, ..., 5\}$, in total. The typical patterns of the three characters are shown in Fig. 5(a).

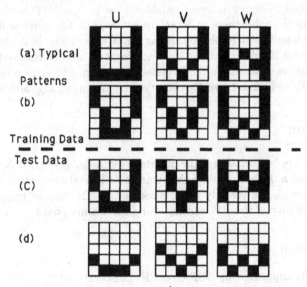

Fig. 5 Character recognition example.

The classification of the input pattern is done by M-neurons in the M-neuron layer. M-neurons for the three characters are directly constructed by *aggregating* the activations of the cells in the typical pattern so that the initial neuron can at least recognize the typical patterns. (Those cells that are not in the typical pattern are connected by 0 weight links because some of them may grow as prohibitive links with negative weights during the BP process). The input pattern is classified to a character that has the largest output value. Note that what matters here is only the order, i.e., which of the three M-neurons outputs is the largest.

The NN using crisp logical functions can only recognize exactly the typical patterns. To give some robustness for noisy and slightly distorted patterns, the fuzzification is necessary: The ideal fuzzified patterns may be such that remain almost invariant under the noise and distortions, but still separable. Further, the continuous property of M-neuron will be useful for the incomplete pattern with some missing pixel information. However, the possibility and how to find such fuzzified patterns by NN needs more consideration and discussion, and it is our future work. Here we concentrate on the illustration of our direct synthesis method using the following fuzzification.

Let's denote the fuzzified input pattern as $\{fn_{ij} = (fx_i, fx_j); i, j = 1, 2, .., 5\}$. First fn_{ij} is constructed as an I-neuron with the single input $n_{ij} = (x_i, y_j)$. Next we connect fn_{ij} to its peripheral cells, i.e., $\{n_{i+1,j}, n_{i-1,j}, n_{i,j+1}, n_{i,j-1}\}$. In this case the weights are determined as one fourth of the weight from n_{ij}. In other words, fn_{ij} is activated to high when either n_{ij} or all peripheral nodes are activated to that level. Thus fn_{ij} is interpreted as "Around n_{ij}."

The initial NN can already distinguish typical patterns of 'u', 'v' and 'w'. The characters in Fig. 5(b),(c), and (d) are recognized as (v, v, u), (v, v, v) and (u, v, u), respectively(Correct recognition should be (u, v, w)). This shows the fuzzification gives the robustness to a certain extent, but the misclassifications occur due to the insufficient resolution. Using 6 patterns in Fig. 5(a) and (b) as the training data for BP, 13 iterations (for each pattern) were necessary to learn correct classifications. (For the correctly classified pattern, the error is set to 0). The NN after BP could also correctly classify the patterns in Fig. 5(c) and (d).

4 Conclusion

We discussed the implementation of the qualitative knowledge into the NN structure. We presented a network builder that translates logical expressions into the NN. By starting with the directly synthesized initial NN, we can improve the efficiency and stability of the BP learning using the training data.

Acknowledgement

Research partially supported by NSF Grant INT91-08632

References

1. D.E.Rumelhart, J.L.McClelland and the PDP Research Group Eds., *Parallel Distributed Processing*, MA:MIT Press, 1986

2. B.Kosko, *Neural Networks And Fuzzy Systems*, NJ: Prentice Hall, 1992

3. R.Serra and G.Zanarini, *Complex Systems and Cognitive Processes*, Heidelberg:Springer-Verlag, 1990

4. Special Issue on Neural Network, *Journal of the Society of Instrument and Control Engineers*, Vol.30, No.4, 1991 (in Japanese)

5. K.Funahashi, "On the Approximate Realization of Continous Mappings by Neural Networks", *Neural Networks*, 2, 183-192, 1989

6. I.Hayashi, H.Nomura, N.Wakami, "Artificial neural network driven fuzzy control and its application to the learning of inverted pendulum system", presented at UFSA '89 Seattle, 610-613, 1989

7. M.M. Gupta, and M.B. Gorzalczny, "Fuzzy neuro-computational techniques and its application to modelling and control", Presented at IFSA'91, Brussels, Proceeding Vol. Artificail Intelligence, 46-49, 1991

8. R.R. Yager, "Using fuzzy logic to build neural networks", Presented at IFSA'91, Brussels, Proceeding Vol. Artificail Intelligence, 210-213, 1991

9. H. Ishibuchi, R.Fujioka, H.Tanaka, "Possibility and necessity data analysis using neural networks", Presented at IFSA'91, Brussels, Proceeding Vol. Artificail Intelligence, 74-77, 1991

10. R.Fujioka, H. Ishibuchi, H.Tanaka, M. Omae, "Learning algorithm of neural networks for interval-valued data", Presented at IFSA'91, Brussels, Proceeding Vol. Artificail Intelligence, 37-40, 1991

Appendix 1 (Proof of Lemma 1)

Let's express w_i as $a_i/(1 - x_i^{Min})$ by using a parameter $a_i > 0$. Because of $w_i(1 - x_i^{Min})$ cannot increase, we can express a_i as

$$a_i = a_m + d_i \qquad (A1.1)$$

where $d_i \geq 0$ and $a_m = Min_{1 \leq i \leq m}\{a_i\}$.
Then we can rewrite (6.2) as

$$w_m(1 - x_m^{Min}) - \sum_{i=1}^{m} w_i(1 - x_i^{Max}) = a_m - \sum_{i=1}^{m} \frac{a_i}{1 - x_i^{Min}}(1 - x_i^{Max})$$

$$= a_m - \sum_{i=1}^{m}(a_m + d_i)\frac{1 - x_i^{Max}}{1 - x_i^{Min}} = a_m(1 - \sum_{i=1}^{m}\frac{1 - x_i^{Max}}{1 - x_i^{Min}}) - \sum_{i=1}^{m} d_i\frac{1 - x_i^{Max}}{1 - x_i^{Min}}$$

$$> \Xi_U^\epsilon - \Xi_L^\epsilon \qquad (A1.2)$$

By moving the term having d_i to the righthand side, we have

$$a_m(1 - \sum_{i=1}^{m}\frac{1 - x_i^{Max}}{1 - x_i^{Min}}) > \sum_{i=1}^{m} d_i\frac{1 - x_i^{Max}}{1 - x_i^{Min}} + \Xi_U^\epsilon - \Xi_L^\epsilon \qquad (A1.3)$$

Because both the righthand side of (A1.3) and a_m are positive, we must have the inequality (7) and, therefore, (7) is necessary.

Now suppose that (7) holds. Given a_m, We obtain $\{w_i\}$ and θ by setting $d_i = 0$ for all i. In this case $w_i(1 - x_i^{Min})$ has the identical value for all i. We can rewrite (A1.3) as

$$a_m > (\Xi_U^\epsilon - \Xi_L^\epsilon)/(1 - \sum_{i=1}^{m} \frac{1 - x_i^{Max}}{1 - x_i^{Min}}) \qquad (A1.4)$$

Let's choose a_m as

$$a_m = \gamma \frac{\Xi_U^\epsilon - \Xi_L^\epsilon}{1 - \sum_{i=1}^{m}(1 - x_i^{Max})/(1 - x_i^{Min})} \qquad (A1.5)$$

where $\gamma > 1$. We use this value for all a_i. Then we have

$$w_i = \frac{a_m}{1 - x_i^{Min}} = \gamma \frac{\Xi_U^\epsilon - \Xi_L^\epsilon}{(1 - x_i^{Min})\{1 - \sum_{i=1}^{m}(1 - x_i^{Max})/(1 - x_i^{Min})\}} \qquad (A1.6)$$

The inequality (6.2) holds because, by plugging (A1.6) into (6.2), we have

$$w_m(1 - x_m^{Min}) - \sum_{i=1}^{m} w_i(1 - x_i^{Max}) = \gamma(\Xi_U^\epsilon - \Xi_L^\epsilon) > \Xi_U^\epsilon - \Xi_L^\epsilon \qquad (A1.7)$$

Thus we find that the inequality (7) is necessary and sufficient for the satisfiability of (6.2).

Appendix 2 (Proof of Lemma2)

Let h_i be x_i^{Min}/x_i^{Max}.
By using h_i, the righthand side of (10) can be written as

$$\sum_{i=1}^{m} w_i x_i^{Max} h_i - \Xi_L^\epsilon = \sum_{i=1}^{m} h_i(w_i x_i^{Max} - \frac{\Xi_L^\epsilon}{\sum_{j=1}^{m} h_j}) \qquad (A2.1)$$

This distributes Ξ_L^ϵ in proportion to h_i.
Similarly the lefthand side of (10) is expressed as

$$\sum_{i=1}^{m} h_i \frac{Min\{w_k x_k^{Max}\} - \Xi_U^\epsilon}{\sum_{j=1}^{m} h_j} \qquad (A2.2)$$

Because (A2.1) is smaller than (A2.2), we have

$$\sum_{i=1}^{m} h_i\{(w_i x_i^{Max} - \frac{\Xi_L^\epsilon}{\sum_{j=1}^{m} h_j}) - \frac{Min\{w_k x_k^{Max}\} - \Xi_U^\epsilon}{\sum_{j=1}^{m} h_j}\} < 0 \qquad (A2.3)$$

Because $h_i > 0$, the above inequality implies that the following inequality must hold for at least one $i = i+$.

$$w_{i+} x_{i+}^{Max} - \frac{\Xi_L^\epsilon}{\sum_{j=1}^{m} h_j} < \frac{Min\{w_k x_k^{Max}\} - \Xi_U^\epsilon}{\sum_{j=1}^{m} h_j} \qquad (A2.4)$$

Because $Min\{w_k x_k^{Max}\} \leq w_{i+} x_{i+}^{Max}$, we have

$$w_{i+} x_{i+}^{Max} - \frac{\Xi_L^\epsilon}{\sum_{j=1}^m h_j} < \frac{w_{i+} x_{i+}^{Max} - \Xi_U^\epsilon}{\sum_{j=1}^m h_j} \qquad (A2.5)$$

and by gathering Ξ_L^ϵ and Ξ_L^ϵ to the righthand side, we can rewrite (A2.5) as

$$w_{i+} x_{i+}^{Max}(1 - \frac{1}{\sum_{j=1}^m h_j}) < \frac{\Xi_L^\epsilon - \Xi_U^\epsilon}{\sum_{j=1}^m h_j} \qquad (A2.6)$$

Because $\Xi_L^\epsilon - \Xi_U^\epsilon < 0$, $w_{i+} > 0$, and $x_{i+}^{Max} > 0$, the following inequality must hold to make the lefthand side of (A2.6) negative.

$$1 - \frac{1}{\sum_{j=1}^m h_j} < 0 \qquad (A2.7)$$

This shows that (11) is necessary.

Suppose that (11) holds. We show the sufficiency of (11) by actually obtaining $\{w_i\}$ and θ that satisfy (10). We choose w_i so that the inequality (A2.6) holds for all i. Because $\sum_{j=1}^m h_j - 1$ is negative, we can have

$$w_i > \frac{\Xi_U^\epsilon - \Xi_L^\epsilon}{x_i^{Max}(1 - \sum_{j=1}^m h_j)} > 0 \qquad (A2.8)$$

Let's decide w_i by using a parameter $\gamma > 1$ as

$$w_i = \gamma \frac{\Xi_U^\epsilon - \Xi_L^\epsilon}{x_i^{Max}(1 - \sum_{j=1}^m h_j)} \qquad (A2.9)$$

This makes all $w_i x_i^{Max}$ an identical value. It is clear that the inequality (10) is satisfied because, by plugging (A2.9) into (10), we have

$$Min\{w_i x_i^{Max}\} - \Xi_U^\epsilon - (\sum_{i=1}^m w_i x_i^{Min} - \Xi_L^\epsilon) = (\gamma - 1)(\Xi_U^\epsilon - \Xi_L^\epsilon) > 0 \qquad (A2.10)$$

Thus we find that the inequality (11) is necessary and sufficient for the satisfiability of (10).

Part IV

Applications of Fuzzy Logic - Control Applications

Fuzzy Control and Neural Networks: Applications for Consumer Products

Noboru Wakami

Central Research Laboratories, Matsushita Electric Industrial Co., Ltd. 3-15,
Yagumo-Nakamachi, Moriguchi, Osaka,670, Japan

Abstract. Fuzzy logic is being implemented in control systems for a
wide range of Japanese consumer products. The reason why fuzzy logic
is used so widely in Japan is described in this chapter. Both the "Fuzzy"
washing machine and the video camera equipped with the "Fuzzy gyro"
are explained. Furthermore, new technology which integrates neural networks and fuzzy logic is reported. This "neuro-fuzzy" technology is already being introduced to the Japanese market.

1 Introduction

1990 in Japan could be considered the first "fuzzy" year. Almost without excep-
tion Japanese appliance and electronic equipment manufacturers strived hard
to develop "fuzzy" appliances and products, and those who succeeded quickly
placed themselves at an advantage in the market. As a result, the word "fuzzy"
became a fashionable new trend in consumer electronics. This trend was surely
triggered by the automatic washing machine called "Aisai (beloved wife) Day
Fuzzy" introduced onto the Japanese market by Matsushita Company in Febru-
ary of 1990. "Aisai" became an unprecedented hit product (see Table 1).

Although substantial advances have been accomplished in the field of home
appliances by employment microcomputers and various sensors, the deployment
of fuzzy control technology realized revolutionary functions which had long been
sought by consumers. In this paper, the concept of fuzzy control developed in
Japanese companies is firstly described, and this is followed by a description of
the typical applications of the technology to several consumer appliances.

products \ company	Matushita Panasonic	Hitachi	Toshiba	Sanyo	Mitsubishi	Sharp	Sony
washing machine	O	O	O	O	O	O	
cloth dryer machine	O		O	O			
vacuum cleaner	O	O	O	O			
rice cooker	O	O	O	O	O		
microwave oven	O	O	O	O	O	O	
refrigerator	O	O		O	O	O	
air conditioner	O	O	O	O			
kerosene heater	O	O		O			
video camera	O			O			
T V				O			O
camera	Cannon , Olympus						
automobile	Mazda , Isuzu						

Table 1. Fuzzy Home Electronics in Japan.

2 Human Electronics

Many electronics companies have developed diversified high-function products satisfying individual needs. Furthermore, the easy-to-use products acheived by human design have been considered essential. From the stand-point of modern technology advancement, home appliances and electronic equipment have to be more humane and "warmly" integrated to our lives. We believe this "Human Electronics" can be realized by combining the advanced technology such as "fuzzy technology" and ergonomic design principles.

The features of fuzzy control technology directly incorporate qualitative expressions, such as the know-how of expert operators, and the "if-then" production rule paradigm. By incorporating fuzzy technology with a man-machine interface, the machine can be brought closer to accommodating its human user rather forcing him to study detailed operating instructions.

3 Application to Home Appliances

Home appliances have to often deal with qualitative data such as "strong" or "weak" - say in heat control - and "a lot of" or "a small amount of dirt" - say in a washine machine. Fuzzy control can accommodate time-honored cooking and washing know-how by embodying the experience of domain experts. At the same time products can be safely used even by the aged or children. The fuzzy logic-based products are operable by simple controls and are capable of accomplishing high degrees of functionality.

3.1 "Fuzzy" Washing Machine

The automatic washing machine "Aisai" on which fuzzy technology was first applied is explained in this section. This "fuzzy washing machine" was publisized through a number of newspaper and TV commercials , and "fuzzy" became quite famous in the mind of the consumer. This washing machine "Aisai" achieved a breakthrough in the function required for automatic washing machines. The machine senses the size of the washing load, the type of detergent being used, and both the quality and quantity of dirty clothes in order to select the most appropriate washing cycle from more than 600 washing cycles.

In the fuzzy washing machine, first, the amount of laundry is determined by the current load on the agitating motor, and the wash is started with both a proper water level and agitation strength according to the amount of laundry. Then, the optical sensor mounted at the drainpipe determines the type of dirt and the degree of soiling by the rate of increasing muddiness and the degree of muddiness respectively[1]. For instance, a rapid increase in muddiness increase is reason to assume that the laundry is soiled by mud, and a slow increase is reason to believe that the laundry is soiled by oil. On the other hand, the high saturated muddiness level is judged to be the higher soiling degree and low judged to be light soiling. The optimum washing mode is then determined by a microcomputer based on these data and the amount of laundry detected by the sensor (see Fig. 1).

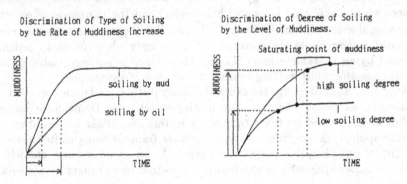

Fig. 1. Changing water clarity patterns detected by optical sensor.

The following are two examples of the fuzzy inference rules: "If the muddiness is *low* and the time to saturate is *short*, then the washing period is set to be *shorter*." "If the muddiness is *high* and the time to saturate is *long*, then the washing period is set to be *longer*".

As a result, excessive washing or inadequate washing can be prevented, and saving of both energy and time can be accomplished[2]. Despite this capability of setting of exact washing cycle, the machine's operation is so simple that it requires no more than a flip of a single button.

As described above, the sensors play highly important roles in the fuzzy logic products and the coordinated development of sensor, microcomputer, and man-machine interface is essential in order to perfect the fuzzy application.

4 Application to Video Equipment

Presently, fuzzy technology is actively applied to TV's and VCR's among video equipment and particularly to video cameras where it has been applied in auto-focusing and auto-iris control. While the video cameras have been used for limited purposes such as recording the growth of children or special events, its mode of use is rapidly shifting toward easier snap shot type use, where the camera is often held by one hand.

Therefore, not only the more compact, light weight, and easier to use video cameras, but also cameras capable of suppressing shaky pictures are strongly desired.

4.1 Video camera equipped with "Fuzzy gyro"

The S-VHS-C type single-hand video camera "Brenby (not shaky)" placed on the market in June 1990 by Matsushita Company is equipped with a "fuzzy gyro" function. The function is realized by both digital video signal processing and application of fuzzy technology. In the video camera, first, the analog video signal is transformed into a digital signal by means of an analogue-digital converter and stored in memory. Secondly, this digital signal is processed to extract the picture shakiness[3].

By applying fuzzy technology, the cause of a shaky picture is discriminated between an unstable hold (so called hand jitter) and the movement of object. Retrieving stored video images is controlled to shift the image along an inverse direction to compensate the movement of image only when the shaky picture is caused by an unstable camera hold[4]. Thus, fuzzy technology makes stable shooting possible while walking or from the window of automobile.

A human can easily judge the cause of a shaky picture qand decide whether it is caused by an unstable hold or a fast moving object in the frame. Such human judgement is replaced by fuzzy technology in this case. That is, according to the rule applied thereto, when an entire picture frame is being shifted toward an identical direction, it is judged to be caused by an unstable hand, and when an object moves differently in the frame, it is judged to be caused by a moving object.

As described above, human levels of judgement become possible by using a microcomputer processor when digital and fuzzy technologies are well combined without using high-degree image processing and recognition technologies, and this is a definite advantage of fuzzy technology.

5 The neuro-fuzzy technology

There are 3 typical methods to integrate neural networks and fuzzy reasoning. We call these neuro-fuzzy technologies.

type 1 a method to apply the fuzzy control and neural networks to different control objectives;

type 2 a method to modify the fuzzy reasoning results by neural networks;

type 3 a method to determine the membership functions of fuzzy rules by neural networks.

In the type 1, neural networks and fuzzy reasoning are applied to different controlled objectives. In the type 2, weighting factors are assigned to each fuzzy inference rule and neural networks are applied to modify the factors using input-output data. As a result, designs of fuzzy inference rules can be roughly defined. In the type 3, neural networks are used to determine membership functions.

In this paper, the type 3 method is explained. When fuzzy reasoning is used for control, the fuzzy control has the problem of tuning. The problem of tuning is how we can determine both the optimal membership functions and the optimal inference rules. In the conventional fuzzy control, human operators determine the membership functions by a heuristically. To solve this problem, some methods in the type 3 are proposed. The following are two typical methods of neuro-fuzzy technologies. One is the NN-driven fuzzy control[5][6] and another is the self-tuning method[7].

In the NN-driven fuzzy control, when input-output data are given, the optimal membership functions are represented by a back propagation type neural network. The optimum membership functions in the antecedent part of fuzzy inference rules are determined by a neural network, while in the consequent parts an amount of reasoning for each rule is determined by other plural networks.

In the self-tuning method, the membership functions in the antecedent part are triangular shape and its consequent part is expressed by a real number. The shape of this membership function and the real number are optimized by a descent method from input-output data gathered from experienced human operators. The descent method is usually used as the learning algorithm for neural network and it is the same as the delta rule of Multi-Layered neural networks.

Although the identification of appropriate control by input and output data is also possible by using the neuro technology, the neuro-fuzzy technology is more advantageous in that there is an easier explanation of the internal controller structure.

By using neuro-fuzzy technologies, fuzzy inference rules can be automatically optimized. In another words, by teaching typical data, not only the taught fact, but also its peripheral knowledge can be optimized using this technology. This is expressed by an aggregation of if-then type fuzzy rules which can be confirmed by designers and experts[8]. This is a highly important matter to the manufacturers who is responsible to offer reliable products to the market and consumers and need to verify appliance behavior.

5.1 "Neuro-Fuzzy" Washing Machine

Neuro-fuzzy technology enables finely tuned and advanced fuzzy control of appliances with a large number of input-output terminals that otherwise would have been impossible to implement using conventional tecniques.

In the "neuro-fuzzy" washing machine, the work performed by an expert consists of unconsciously controlling the washing and rinsing cycle. This is done not only by considering the degree of soiling and volume of laundry but also the type of water and dissolution rate of soap. Sensed data of this type are used to construct an if-then rule control regime and the fuzzy inference process is optimized w.r.t. fuzzy set membership functions. The result is a finer degree of control of wash time can be achieved and the wash cycle can be executed by discriminating on finer inputs not known by the human operator, such as the degree of soap dissolution The result is that the neuro-fuzzy washing machine can select the most appropriate washing cycle from a choice of more than 3,800 possibilities.

6 Conclusion

The "intelligent" capability of home appliances and electronic equipment which have been drastically advanced by using both a microcomputer and various sensors are now producing more user-friendly products through applying fuzzy technology.

As described above, the further development of "Human Electronics" is highly important. For this, the developments of flexible artificial intelligence centering around fuzzy technology, fusion with neurocomputers, and human interface for the coordination of human and machine are essential. By accomplishing these, the incorporation of human judgement, behavior, and concept into the machine and system would become possible. In this movement, the application of fuzzy technology to the home appliances and electronic equipment will be more extensive, and this could be applied further to machines which could understand the human language. The application of these possibilities to a robot with an advanced learning function may revolutionize society.

References

1. Yamashita, H., Abe, H.,Kondo, S.,Kiuchi, M. and Imahashi, H.: A Washing Process Monitoring Sensor for Fully Automatic Washing Machine Using Fuzzy Inference. 9th Sensor Symposium, pp. 163-166 (1990).
2. Kondo, S., Abe, S., Tarai, H., Kiuchi, M. and Imahashi, H.: Fuzzy Logic Controlled Washing Machine. Proc. of IFSA'91 Brussels, Engineering pp.97-100, (1991).
3. Uomori, K., Morimura, A., Ishii, H.,Sakaguchi, T. and Kitamura, Y.: Automatic Image Stabilizing System by Fully-Digital Signal Processing. IEEE Trans. Consumer Electronics, Vol. 36, No. 3, p.510, (1990).
4. Egusa, Y., Morimura, A., Akahori, H, and Wakami, N.: An Electronic Video Camera Image Stabilizer Operated on Fuzzy Set Theory. Proc.of IEEE International Conference on Fuzzy Systems (FUZZ-IEEE'92), (1992).
5. Takagi, H. and Hayashi, I.: NN-Driven Fuzzy Reasoning. IJAR, vol.5, pp.191-212, (1990)

6. Hayashi, I., Nomura, H.and Wakami, N.: Learning Control of an Inverted Pendulum System by Neural Network Driven Fuzzy Reasoning. Proc. of the Second Joint Technology Workshop on Neural Networks and Fuzzy Logic, NASA Conference Publication 10061, vol.l, pp.169-182, (1991).
7. Nomura, H., Hayashi, I. and Wakami, N.: A Self-Tuning Method of Fuzzy Control by Descent Method. IFSA'91 Brussels, Engineering pp.155-159, (1991).
8. Nomura, H., Hayashi, I. and Wakami, N. : A Learning Method of Fuzzy Inference Rules Descent Method. Proc. of IEEE International Conference on Fuzzy Systems (FUZZ-IEEE'92), (1992).

Part V

Applications of Fuzzy Logic - Fuzzy Logic in Planning

Case-based Reasoning for Action Planning by Representing Situations at the Abstract Layers

Toshihiko Yokogawa[1] and Takefumi Sakurai[2] and Akira Nukuzuma[1] Tomohiro
Takagi[1] and Shigenobu Kobayashi[4]

[1] Kayaba Industry Co., Ltd. 1-12-1 Asamizodai, Sagamihara-City, Kanagawa, 228
Japan
[2] Laboratory for International Fuzzy Engineering Research, Sieber Hegner Building
4F, 89-1 Yamashita-cho, Naka-ku, Yokohama, 231 Japan
[3] Laboratory for International Fuzzy Engineering Research, Sieber Hegner Building
4F, 89-1 Yamashita-cho, Naka-ku, Yokohama, 231 Japan
[4] Tokyo Institute of Technology, 4259 Nagatsuta, Midori-ku, Yokohama, 227 Japan

Abstract. A robot which can act according to natural language instructions must complement information from experiences and common
knowledge to adapt to various situations. In this paper, we propose a
method to evaluate the similarities of concepts by representing situations
in abstract layers which are hierarchically arranged. We also present a
method for case-based reasoning for planning which uses the layers as
viewpoints for case retrieval. The technique is applied to the action planning for the house-work robot.
Keywords: Action Planning, Case-Based Reasoning, Multi-layer Representation, Measurement of Similarity.

1 Introduction

One "human-friendly" system is a human-machine communication system which
receives language commands from users, understands them, and makes a plan
with appropriate command sequences for a robot. The assumed robot is autonomous, i.e. moving itself and not working in a fixed location. The problem
with such an idea is that linguistic commands have an inherent vagueness, ambiguity, and variety of meanings. Ambiguity and vagueness are inevitable in
natural language expressions. Ambiguity or multiplicity of meaning is dependent on context; for example, the agent who issues the command, the situation
in which it is given, and so on. In other words, a command sequence does not
always have the same unique meaning or request for the same action sequence.
What a user wants the robot to do is determined by the situation. To understand
language commands with such inherent ambiguity, dependent on contexts and
situations, the utilization of past experience may be useful. Valid use of past
experience is performed based on Case-based reasoning.

To understand a command with compound elements and plurality of meaning, flexible reasoning according to the present situation is needed. Situations
must be interpreted differently depending on the purpose of the reasoning and

this must be done during understanding and planning. Viewpoints for interpretation must be set according to the level and the process of reasoning. They are typically defined by the level of abstraction. The reasoning proceeds from an abstract level or viewpoint to the more concrete level or viewpoint.

2 Directions of the Study

One of the main problems in case-based reasoning is how to retrieve the best-fit case from the "case base" [A database of past cases and problem solving experiences]. Usually, the retrieval is done as follows; features of cases and the corresponding problem are represented by attribute-value or slot-value pairs and matching is done by calculating how many attributes the cases and corresponding problem have in common. A quantitative measure of the "best match" determines the best matched case. In some examples, the subset of attribute-value pairs are used to index cases and the rest of attribute-value pairs used for whole matching. In other systems some weights are set to certain attributes [2, 7, 3].

A straightforward approach to case retrieval from fuzzy theory is to add fuzzy labels and concepts to the value part of the attribute pairs instead of crisp values and/or concepts[10].

In order to apply case-based reasoning to the action planning in the cleaning robot example, differences when reasoning with abstraction levels must be considered. It depends on the purpose of the reasoning in the total process of planning. For example, when attempting to achieve a general or macroscopic level of target actions, the reasoning must be done in the abstract level; when trying to achieve the order of target actions, the reasoning is performed by paying attention to combinations of the objects; and when the goal is concrete actions and its features, more concrete levels or viewpoints must be attended to. In other words, these differing situations must be understood from different viewpoints according to the level of reasoning or abstraction level of the target reasoning. For example, a situation like "some coffee is spilled on a table" is represented as "(spilled coffee[quantity= ..] on-table)" in the concrete level and used to decide concrete actions and features for such actions. On the other hand, the same situation is interpreted differently in "(exist DIRT on-WorkingPlace)" and used to decide the general purpose of the action requested by the user.

When concepts or objects are set to the value part of the attribute of a case or problem, the matching degree, based on similarity, is calculated. Similarity between two concepts is usually decided by whether they are descendant nodes of the same node and the degree is calculated by the distance in the conceptual tree and matching degree of attribute-value pairs of the concepts[5, 7]. The similarity between two concepts in different situations however cannot be calculated uniquely because of the different viewpoints or levels of abstraction. For example, "coffee" is very similar to "tea" at the concrete level of thinking. At the abstract level, on the contrary, "coffee (which is spilled on table)" and "tea (which is in the teacup)" are not as similar as the individual concepts "coffee" and "tea".

To evaluate such differences (and similarities) between the two concepts, we introduce abstract layers reflecting the levels of abstraction. These provide the appropriate viewpoints for situation and subsequent. Situation and individual concepts are represented at these layers. Case retrieval is performed with the layer level and one or more cases may be selected in any given level. The approach in this paper is based on the idea that a "fuzzy approach" involves macroscopic identification of situations where the evaluation of the multiplicities of concepts and meanings can be performed.

3 Representation of the Situations and Concepts at the Abstract Layers

3.1 Setting of the Layers

The abstract layers are stratified fields where reasoning with a certain goal proceeds and situations and concepts are represented by the abstraction level appropriate to the goal. These reflect viewpoints for the reasoning goal. Represented concepts and situations in layers is analogous to the interpretation of the real objects and situations from the layer's viewpoint. Establishing changes to these layers based on the goal of reasoning represents the situations that the planning process has interpreted.

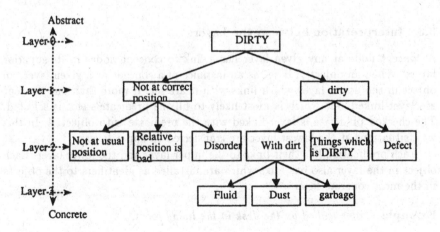

Fig. 1 A Tree of Cognition for the Layer Setting.

In the example of planning a cleaning robot, the layers are set based on the tree-structure of the recognition of situations of the status "dirty". Each horizontal level of the tree at Fig. 1 is set to each layer. The highest level (the level of node "DIRTY") in Fig. 1 is set to Layer 0, which is the most abstract layer and level of target field description. The second level (the level of "Not at correct position") establishes a layer where reasoning is done to extract the main goals and their general orders in the target field. The lowest level (the level of "Fluid") is set to the most concrete layer to represent real situations, where the reasoning is done to extract real action commands.

3.2 Descriptions at the Layer

Each layer has its own vocabulary (list of concepts and predicates) and structure. Concepts are not necessarily pure ones, but they represent some kind of interpretation at the layer of concrete concepts. They are constructed as a conceptual tree or network which means interpretational parent-child relations. Predicates have slots, the kind of relations the slot fillers have when the predicate is initiated, and likelihood that concepts can fill the slots. A slot value is usually an instance of a concept in the same layer.

Situations and actions at each layer are represented by a list of predicates and its filler for each slot. Objects are represented as the instances for concepts and have some associated attributes. What kind of attribute and what kind of value are restrained by the layer in accordance with the layer's abstraction level. In the concrete layer, the kind of attributes are concrete i.e., "length," "weight", and attribute values may be numeric values. The upper layer has the same attributes but the values allowed are fuzzy labels and the upper-most layer has no such attributes.

3.3 Interpretation between the Layers

A concept node at any given layer has a link to concept nodes in its superior layers. When an object A is set as an instance of a concept at a given layer, an object in the upper layer which links with it are set. If more than one concept are layer-linked, one which is most likely to fill the predicate's slot is selected. The checked predicate is layer-linked with the predicate with object A. In this way, situations at the lowest layer are transferred to upper layers.

Each predication has their links to the upper layer and the lower layer. Each object in the layer also has links which are installed as identifiers to the objects in the most concrete layer.

Example *Coffee spilled on the desk in the living room.*

```
(Layer 1              <- abstract layer
    ((in WorkingTable1 LivingRoom2)
        ((on THING3 WorkingTable1) (DIRTY WorkingTable1))))
(Layer 2              <- medium layer
    ((in DESK1 LivingRoom2)
        (spill  FLUID3 on-DESK1)))
```

```
(Layer 3                    <- concrete layer
   ((in DESK1 LivingRoom2)
    (spill  COFFEE3 on-DESK1)))
```

3.4 Similarity between Concepts in Layers

Equation (1) represents the similarity of the two concepts A and B in layer n.

$$DL_n(A, B) = \sum_{i=-1}^{m} W(n + i) \times D_{n+i}(A, B) \tag{1}$$

where DL_n: Integrated similarity at layer n.
$W(n)$: A coefficient which measures the weight of layer n.
D_n: Similarity of concepts within layer n.

When a layer is addressed, similarity is calculated not only by the similarity within it but also by that in several of the layers around it. By reducing relative weights at the layers immediately adjacent (contributions for the similarity of the layers adjacent), a membership function is defined. This means a viewpoint can be weighted (Fig. 2). Adjusting the membership function makes it possible to give weights which are appropriate to reasoning goals.

D_n (Similarity of the two concepts) is calculated from the nearness between two concepts in the concept tree within the layer (Equation (2)).

$$D_n(A, B) = \frac{1}{(M(M) - L(A)) + (L(M) - L(B)) + 1} \tag{2}$$

where M: The ancestor node common with concept A and B in conceptual tree[5]
$L(concept)$: level number of concept in the tree.

When the concepts have properties, matching degrees of the kinds of properties and their values can be easily integrated.

3.5 Similarity between Predications at the Layer

Similarity between predications PA and PB ($DL_n(PA, PB)$) is calculated from the similarity of predicates of them (PredA and PredB) and the objects at the slots of the predicate (ClauseA and ClauseB) as Equation (3)

$$DL_n(PA, PB) = f(n)*DL_n(PredA, PredB)+(1-f(n))*Ave(DL_n(clauseA, clauseB)) \tag{3}$$

$f(n)$ is adjusted to set relative weight to the predicate match according to the abstraction level in layer n.

---

[5] If an ancestor node common with A and B does not exist, $D_n(A, B) = 0$.

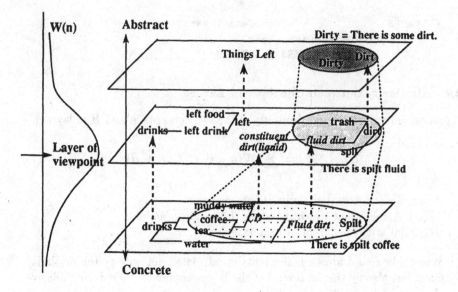

Fig. 2 Description of a situation at the abstract layers and weighting
membership function.

4   Case-Based Reasoning Using the Abstract Layers

4.1   Outline of the Process

We explain a method of case-based reasoning which solves problems at the abstract layer level for the given planning goal. Fig. 3 shows the general flow of the case-based reasoning using the abstract layers.

1. **Assignment of a problem to an appropriate layer.**
   Decide the abstract layer where the first problem solving takes place and assign the problem to the layer. In the example of action planning by the language instructions, the layer is assigned in accordance with the level of abstraction in the expressions.

2. **Retrieval of cases at the assigned layer.**
   Retrieve cases in the assigned layer. Situations[6] for the problem and cases are compared in the layer and similarities calculated. Calculation of similarities is described in 4.2.

3. **Revising of the retrieved case at the assigned layer.**
   The results of the best-matched case in the current layer are revised to fit the situations relevant to the problem. The revision is done by replacing matched pairs of concepts.

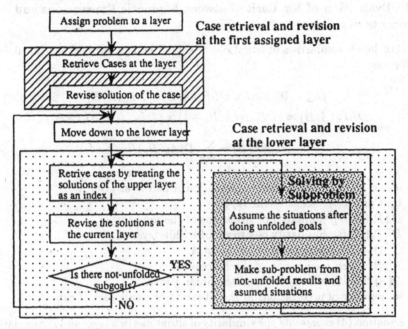

Fig. 3 General flow of case retrieval and case revision in the case-based reasoning using the abstract layers

4. **Unfolding the results of a layer at the lower layer.**
   After finding results in a layer, move down the current layer by one class and retrieve cases to find results in it. That means unfolding the results of the upper layer at the lower layer. The results of the upper layer are used as indices to limit the area of cases searched. A partial list from the first of the results at the upper layer must be matched before matching the situations at the current layer. In this way, some goals of the result in the upper layer are developed to more concrete goals at the level of the current layer.

5. **Treatment of the not-unfolded goals.**
   Not-unfolded goals of the result in the upper layer are generated as the result of a sub-problem and resolved at the current layer. The sub-problem has situations which are based on the assumption that the actions of unfolded

---

[6] Cases and problems are represented by the situation part and result part at each layer. Each part is a list of predicates and link information.

goals at the current layer are complete. If the not-unfolded goal remains at the upper layer, this sub-problem method is continuously executed. Because assumed situations have no details at lower layers, sub-problems are not able to unfold at the current stage. A depth-first method of reasoning is therefore applied; earlier goals are unfolded to the most concrete level first and, after that, the remaining goals are unfolded from the layer where sub-problems branch off.

4.2 Evaluation of Similarity between Semantic Expressions and Concepts at a Layer

At each layer, similarities of situations of case-A and problem-B are calculated as follows;

$$D_n(A, B) = k_p \times DLP_n(A, B) + DL_n(A, B) \tag{4}$$

$$DLP_n(A, B) = (DL_{n-1}(A, B) + 1) \times (DL_{n-1}(G(A), G(B))) \tag{5}$$

$$DL_n(A, B) = \sum_{h=1}^{m} D(A_h, B_h)/m \tag{6}$$

where $D_n(A, B)$: Similarity of situation A and B at layer n.
k_p: Constant belongs to the layer.
$DLP_n(A, B)$: Similarity of situations at the upper layer.
$DLP_n(A, B)$: Similarity of situations within layer n.
$G(A)$: Goal descriptions of situation A.
A_i, B_i: Each predication of the description.
$D(A_i, B_i)$: Similarity between the predications.
m: number of predications in the situation or result part of case-A.

Equation (4) represents the similarity of situations in a layer and is calculated by the similarity at the current layer and that at the upper layer with some weight. This means a similarity at the more abstract level is considered when calculating similarity to unfold the results at the upper layer in more concrete ones. Equation (5) shows that the similarity of the situations at the upper layer is calculated by integrating similarities of the situation part and resultant. This also means that when a resultant does not exist,[7] the similarity of the situations in a layer are calculated by the similarity within the current layer.

Equation (6) expresses that similarity within the layer is the average of similarities of the predications of case-A in that layer. When the situation part has several predications, a comparison is made between the predications which have greater similarity to a predicate and have the same number of terms. Similarity of predications in the layer is calculated by equation (3) (See 3.5). In the example of language instruction understanding for the house-work robot, the similarity between two predicates used in the predications is assigned $(1, 0)$ from whether they are identical or not.

---

[7] Step 2) at the flow. (See 4.1).

5 Application to the Action Planning of Cleaning

We show an example of the method of case-based reasoning described in section
3. This is applied to language instruction understanding for an autonomous robot
for in-house cleaning. A problem is as follows:

instruction: "Clean the desk, because it is dirty."
situation: Some coffee spilled on the desk in the child's room.
There is an ornament made by glass on it.

1. **Assign the first layer to the layer 1 from the expression of reason
in the instruction.**

```
[Description of PROBLEM]
(Instruction (Reason "with dirt") (Object  Desk1)
      (Movement clean) (Instructor Housewife))
(Layer 3
(((in  Desk1 Child'sRoom2) (spill Coffee3 on-Desk1))
  ((on  Ornament4 Desk1)  (Made-by Ornament4 Glass))))
(Layer 2
(((in  Desk1  Child'sRoom2) (spill  Fluid3  on-Desk1))
  ((at  Ornament4  position) (Made-by Ornament4 Glass))))
(Layer 1
(((in  WorkingTable1 Child'sRoom2) ((on Dirt3 WorkingTable1)
      (Dirty on-WorkingTable1))
  ((at  Thing4  position)))))
```

2. **Retrieve cases by the situations of the Layer 1.**
   In this retrieval, similarity of situations between a problem and each case is
   calculated by dividing the situations into each predication and make pairs
   which have a same predicate.

   > **[CASE-1]**
   > **Instruction:** "Clean the desk."
   > **Situations:** There is a cup on the desk. Some coffee is left at the
   > cup.

```
(Layer 1 (Goals
((MOVE Thing3 (from WorkingTable1) (to WashingFasitly)))
(((in  WorkingTable1 LivingRoom2)
¨(at Thing3 position) (in Left4 Thing3))))
```

In CASE-1, the "cleaning" is understood as "clean the things away." CASE-1
and the Problem have in common the existences of the similar things at the
concrete level. At the abstract layer where the reasoning begins, however,
they do not match.

[CASE-2]
Instruction: "Clean the floor because it is dirty."
Situations: There is dust and trash on the floor of the living room.
There is a chair on the floor.

```
(Layer 1 (Goals
((MOVE Thing5 (from WorkingPlace1) (to TemporaryPlace))
 (MOVE-OUT Dirt3,4 on-Floor1))
 (((in WorkingPlace1 LivingRoom2)
  ((on Dirt3,4 WorkingPlace1)(Dirty on-WorkingPlace1))
  ((at Thing5 position))))
```

The situations are divided into each predication and matching is done between the predication which have the same predicate in common. For example; (in WorkingTable1 Child'sRoom2) is tried to match (in WorkingPlace1 LivingRoom2). In this matching, the object "WorkingTable1" matches "WorkingPlace1" and "Child'sRoom2" matches "LivingRoom2[8] The matching of these two predications, therefore, succeeds with some degree of similarity. In this way, the object "Dirt3" and "Dirt3,4" matches and so on; all predications have their matched predications; and the whole situations match at a certain matching degree and the synthetic degree of similarity is calculated. The calculation is done based on equation (4) - (6). When the degree of matching with CASE-2 is highest, it is retrieved as a matched case. At the same time, matched object pairs are saved as a matched object list as follows; ((WorkingTable1 WorkingPlace2) (Chid'sRoom2 LivingRoom2) (Dirt3 Dirt3,4) ....)

3. **Revise the results of CASE-2. (Fig. 4)**
   The result part of CASE-2 is revised by replacing objects by matched object list obtained by step 2. For example, "WorkingPlace1" is replaced to "WorkingTable1" The objects which does not exist in the situations of the case are untouched in the results. The results at layer 1 are, therefore, as follows:

```
(Layer 1   (Goals
((MOVE Dirt3 (from Here) (to TemporaryPlace7))
 (MOVE-OUT Thing4 on-WorkingTable1)))
```

In the result of the case retrieval at layer 1, the meaning of "clean" in the instruction is understood not as "clean the things away," but as "clean out the dirty" by CASE-2 at the abstract level.

4. **Move into the lower layer (Layer 2).**
5. **Retrieve cases by the result of the upper layer(Layer 1).**
   In this step, the cases in case base are divided in two groups. Only the cases which match in this stage are treated as the objects of the case retrieval by the situations at the current layer.

---

[8] At Layer 1, there are concept hierarchies as follows:
(WorkingPlace IS-A PLACE) (WorkingTable IS-A WorkingPlace) (LivingRoom IS-A Room) (Child'sRoom IS-A Room) etc.

6. **Retrieve the best-matched case by the situations of the current layer from the cases retrieved by step 5).**
In this case, the first two goals of the upper layer and the situations are matched to CASE-3.

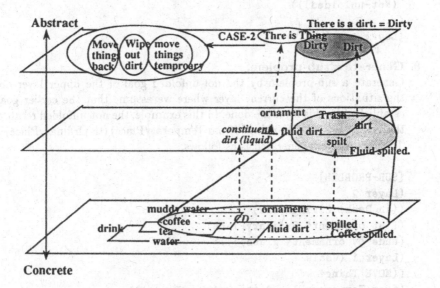

Fig. 4 Case-based reasoning using the abstract layers (1) Retrieve a case and revise it at the first layer

[CASE-3]
Instruction: "Mop the corridor."
Situations: Water spilled on the corridor.
          There is a chair on the corridor.

```
(Layer 2 (Goals
((MOVE Chair3
(from on-corridor1) (to OrdinaryPlace))
 (Wipe on-corridor1 with-Tool5)))
((Spill Fluid2 on-WorkingPlace1)
(on chair3 WorkingPlace1)))

(Layer 1 (Goals
((MOVE Thing3 (from Here) (to OrdinaryPlace))
 (MOVE-OUT Dirt2 on-WorkingSpace1)))
( ....
```

7. **Revise the results of CASE-3. (Fig. 5)**
The objects matched are replaced and the results of the current layer of the problem are as follows:

```
(Layer 2 (Goals
((MOVE Ornament4
(from on-Desk1) (to TemporaryPlace))
 (Wipe on-Desk1 with-Tool5)
 (Not-unfolded)))
(.....
(Layer 1   (....
```

8. **Generate a sub-problem.**

Generate a sub-problem by the not-unfolded goal of the upper layer and the situations of the current layer where we assume that the earlier goals established by this layer are done. In this example, the not-unfolded result at the layer 1 is ((MOVE Thing4 (from TemporaryPlace) (to OrdinaryPlace))). The sub-problem generated is as follows;

```
[SUB-PROBLEM]
(Layer 2
((in Desk1 LivingRoom2)
(("at ornament4 OrdinaryPlace)
(Made-by ornament4 glass))))
(Layer 1 (Goals
((MOVE Thing4
(from TemporaryPlace) (to OrdinaryPlace)))
```

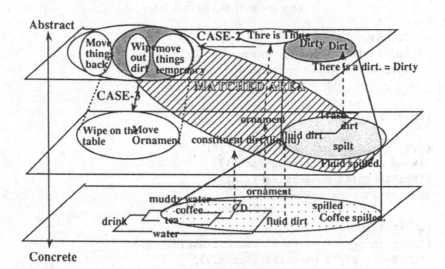

Fig. 5 Case-based reasoning using the abstract layers(2) Retrieve a case at the lower layer.

9. Retrieve cases by a sub-problem. (Fig. 6)

Resolving the sub-problem is done as an ordinary problem by the retrieval and revision of the other cases. Continue these processes until there is no not-unfolded goal of the upper layer.

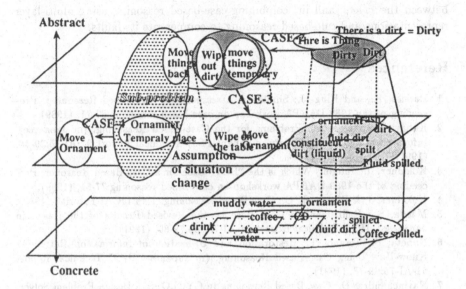

Fig. 6 Case-based reasoning using the abstract layers (3) Generate a
sub-problem and unfond it.

6 Conclusion

To understand language instructions correctly and to make a plan of actions required, the use of experiences in similar situations may be useful. Similarities between concepts in various situations are not constantly evaluated as being of the same degree in different stages of reasoning and in different situations; the similarity must be evaluated w.r.t. the abstraction level set by the goal of the reasoner.

In this paper, we proposed abstract layers which reflect the reasoning goals as a viewpoint. We also proposed a method for evaluating similarity in abstract layers to interpret situations appropriate to a certain reasoning goal. This approach was applied using case-based reasoning for language instruction understanding and action planning for a housework robot and demonstrated the validity of the method. In the method we described, the calculation of similarity between concepts and situations can be based on the goal of reasoning. This can be done since the layers are set to reflect viewpoints to interpret situations, the representations at each layer are restricted for the goal, and similarities are calculated based on the layers. Case-based reasoning by this method can grasp multiple interpretations of the same objects by representing them at the layers and by inferring from abstract level to concrete level. The application of the idea to language instruction understanding can appropriately treat the ambiguity dependent on situations by grasping situation multiplicity.

Future works includes studies; i) how to set the abstract layers appropriately to a field of problems; ii) how to make interpretations and links correctly between the layers, and iii) combining case-based reasoning using multi-layer representations and rule-based reasoning to compensate its faults.

References

1. Bareiss, R., and King, J.: Similarity Assessment in Case-based Reasoning. Proceeding of the 1989 DARPA workshop on case-based reasoning:67-71, (1989).
2. Kobayashi, Shigenobu,: Problems for the Case-Based Reasoning. (in Japanese), Information Processing Society of Japan(IPSJ) Technical Report 91-AI-75:29-38, (1991).
3. Kolodner, J.: Judging Which is the "Best" Case for a Case-Based Reasoner. Proceeding of the 1989 DARPA workshop on case-based reasoning:77-81, (1989).
4. Kolodner, J. & Riesebeck, C.: Case-Based Reasoning. 11th IJCAI Tutorial, (1989).
5. Maeda, S.,: Refinement of Case Base by the Revised Results of the Case. (in Japanese), IPSJ Technical Report 91-AI-75, pp.79-85, (1991).
6. Nabeta, S., Terano, T.: Restoration and Utilization of Information Retrieving Know-How Using Case-Based Reasoning. (in Japanese), IPSJ Technical Report 91-AI-75:69-78, (1991).
7. Navinchandra, D.: Case Based Reasoning in CYCLOPS, a Design Problem Solver. Kolondner, J (ed) Proceeding of a Workshop on Case-Based Reasoning:86-301, (1988).
8. Proter, B.: Similarity Assessment: "Computation vs. Representation". Proceeding of the 1989 DARPA workshop on case-based reasoning:82-84, (1989).
9. Thagard, P., Holyoak, K.: How to Compute Semantic Similarity. Proceeding of the 1989 DARPA workshop on case-based reasoning:84-86, (1989).
10. Zhang, W., Sugeno, M.: A Dynamic Memory Model for Scene Understanding. Proceeding of 6th Fuzzy System Symposium (Tokyo):211-214, (1990).

A Unified Formalism for Landmark Based Representation of Maps and Navigation Plans

Svetha Venkatesh and Dorota Kieronska

School of Computing Science, Curtin University of Technology, PO Box U1987, Perth, Western Australia, svetha@cutmcvax.cs.curtin.edu.au, dorota@cutmcvax.cs.curtin.edu.au

Abstract. We present a unified formalism for representing maps and using them for constructing plans of navigation for an autonomous agent. The foundation of this work lies in addressing key questions that an agent is confronted with when navigating. That is, besides the main task of how to reach the intended destination from the current position, the agent faces other questions like: where am I? what landmarks can I see? where is my destination relative to me and the landmarks I am seeing? Fundamental to this representation is the use of visual landmarks, which are used as pivotal points in the landscape being described. Further, in the representation of spatial information and navigation there are three different viewpoints: first, the localized representation from the viewpoint of a sighted, mobile agent; second, the static representation seen by the map-maker; and third, the view of an external agent giving directions on the basis of his own experience/knowledge. The major contribution of this map model and the associated navigation method lies in the framework which unifies these three different points of view. This unification enables the agent to make no distinction in terms of following implicit instructions contained in a map and the directions given by external agents.

1 Introduction

One of the most common tasks performed on a day-to-day basis by most of us is travelling from one location to another, be it from home to work, from one office to another or just around the house. We are able to do so with little exact information; in fact, estimates of distances, relative direction and position, and orientation with respect to known landmarks are all that we need. When travelling in an unfamiliar environment, we are capable of reading a map and interpreting the information contained therein with respect to observed visual input.

A great deal of effort has been directed towards the building of intelligent systems that can cope with dynamic environments. This large area of research has addressed the problems of robot exploration and navigation in large-scale spatial environments. The representation of spatial environments forms a fundamental aspect of systems that can cope with dynamic worlds. In this regard, two subproblems can be formulated. First, the representation of an unexplored, unmapped terrain, and second, the representation of mapped terrain. In this paper we address the issues of representing maps, for the purpose of map construction and navigation. This method can also be applied to the problem of following another agent's directions.

Levitt and Lawton levitt:90 have developed a qualitative navigation system for a mobile robot in an unstructured environment. Their system has no memory, supports perceptually based navigation and is capable of navigation and guidance in large scale space (i.e. not limited to path following) using purely local visual information.

We present a unified formalism for representing maps and using them for constructing plans of navigation for an autonomous agent. The foundation of this work lies in addressing key questions that an agent is confronted with when navigating. That is, besides the main task of how to reach the intended destination from the current position, the agent faces other questions like: where am I? what landmarks can I see? where is my destination relative to me and the landmarks I am seeing?

Fundamental to this representation is the use of visual landmarks, which are used as pivotal points in the landscape being described. Further, in the representation of spatial information and navigation there are three different viewpoints: first, the localized representation from the viewpoint of a sighted, mobile agent; second, the static representation seen by the map-maker; and third, the view of an external agent giving directions on the basis of his own experience/knowledge. The major contribution of this map model and the associated navigation method lies in the framework which unifies these three different points of view. This unification enables the agent to make no distinction in terms of following implicit instructions contained in a map and the directions given by external agents.

Kender & Leff[2] refer to the map of the geographical terrain as a 'world model', and the 'custom map' as the list of directions to get from one place to another. In the formalism we develop, the augmented map is termed the world map - and can be described in terms of a list of directions and landmarks. The agent plan - can be derived automatically from the world map and that serves as directions to go from one place to another.

The strengths of our formalism are fourfold. Firstly, we can represent complex environments which are not restricted to linear cases, such as corridors of a building. Unstructured environments not containing any marked paths, but instead purely based on landmarks can be expressed, as well as mixed environments, for example in a factory where there can be a net of paths and corridors linking storage areas which are largely unstructured. Secondly, the representation is compact and identical from both the agent's and the map-maker's points of view. Thirdly, the techniques are robust enough to allow for navigation even if intermediate landmarks are obscured or missing. Lastly, our formalism provides the foundation for an autonomous system capable of directly accepting directions from a person.

The outline of this paper is as follows: Section 2 details our proposal for establishing the position of the landmarks. Section 3 outlines how the map can be constructed using these landmarks, from both the agent's and map-maker's points of view. The discussion follows in Section 4 and the conclusions in Section 5.

2 Establishing the position of landmarks

Selectivity is one of the key aids to human memory. When humans walk along a route, they memorize the most distinctive objects or characteristics of scenes. While the quantitative description of distinctiveness is difficult, attempts are being made to describe objects that are distinctly different by qualitative means [KuB90, TsZ90]. In this paper, we refer to such distinctive objects as landmarks. The description of the process of landmark recognition is beyond the scope of the present paper. Instead, we concentrate on the general formalism for map representation, given an arbitrary set of landmarks which are assumed to be within the recognition capabilities of the sensory controller and processor.

In a map representation in terms of landmarks, the viewpoint could be described either in terms of relative or absolute position of landmarks. We propose that both the map-maker and the navigating agent share the same viewpoint, and for this to be possible, the viewpoint must be independent of viewing direction. Towards this end, we suggest that while making maps or navigating, the directions are expressed in terms of readings from a compass. We make no assumptions about the method of constructing the map. It can be done via exploration [3] or by a human.

One of the problems with autonomous agents that navigate using a fixed map, and relative positions with respect to an initial starting position is that errors tend to accumulate. One effective way of dealing with this problem is to reposition the agent at regular intervals, using robust repositioning methods. Visual landmarks can be used to provide robust spatial clues, and we use these clues to constantly assess agent's location.

A further issue that arises when talking about landmarks is the establishment of the landmark position relative to the agent. An agent can arrive at the landmark in different ways depending on the situation. It can use a path to get to a destination, if such a path exists, or in open terrain, follow the straight line line-of-sight path to the destination. In either case, the direction of the navigating path needs to be represented. Exact angles can be used to do this, but in a real world situation, this method will run into problems very quickly because of the noise and inaccuracies of the sensor. A more robust mechanism would be to specify a possible range of angles where the path could lie, whereby this range of angles can serve as the guiding factor to the gaze control mechanism of the agent, and the vision system can then look of the attributes of the path in the image obtained. The agent in this case can tolerate fairly large inaccuracies in the gaze control angle, for as long as the agent looks approximately in the right direction, there is a high likelihood of the agent finding what it is looking for.

Thus, varying road widths and the inaccuracy of measurement, the use of exact angles is impractical and unjustified, a fuzzy set [Zad81] representation is considered more appropriate.

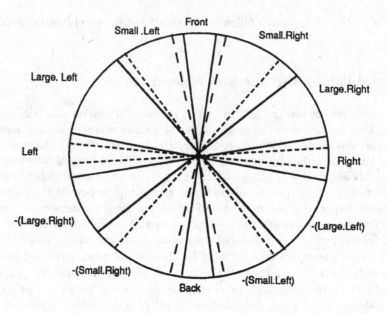

Fig. 1

In Fig. 1, each sector represents an approximate direction. Since the concepts of front, back, left and right have an intuitively clear local interpretation, we choose to use fuzzy sets, named accordingly, for each of these directions. The intermediate regions are divided into further regions based on a direction defined in terms of magnitude and relative deviation from North, and specified as: Magnitude.deviation, e.g. Small.Right. A corresponding division applies in the opposite direction, South. To minimize the terminology, we use a minus sign '-' to indicate a reverse orientation. The fuzzy set membership function for the first quadrant is shown in Fig. 2. The membership functions for the remaining quadrants are defined in an analogous manner.

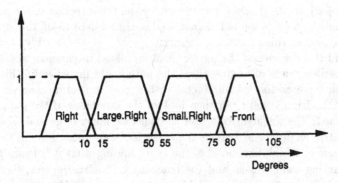

Fig. 2

3   The agent's and map-maker's description of the map

Fundamental to the description of the map are two features: landmarks and landmark relative positions. In this section, we first describe models that can be developed to describe maps in terms of those landmarks and how agents can use them for navigation.

3.1   Map maker's point of view: the world map

Whilst the navigation plan derived for the autonomous agent to go from one spatial point to another can be viewed as a dynamic one, the map-maker views the geographical terrain as a static spatial layout. However, the underlying static model of the map-maker's can be translated to the dynamic model of the agent's, and it is this underlying fundamental dependency that we seek to exploit in our models.

From the map-maker's point of view, landmarks are viewed as spatial situations, and the position of one landmark with respect to another is viewed as a static event that defines how to get from one landmark to another. The situation is further complicated as each landmark can be viewed from different directions, giving rise to different viewing scenarios. For example, consider the landmark 4, in Fig. 3a. Depending on which side of the landmark the agent is, different further landmarks are visible, and in this case, whereas from position 4 landmark 3 is visible, from position 5 landmark 5 is visible. Both positions, 4 and 5, are different viewing positions of landmark 4. From position 4, landmark 4 is positioned on -(Large.Left), and from position 5, landmark 4 is positioned on (Large.Right). This information must also be incorporated in the map.

We use a directed graph to represent a map. The nodes are associated with landmarks and their position from the point of view of the agent. We term each of these nodes viewing points. The arcs represent the direction of the path to be followed from one landmark to the other. Both, the position of the viewpoint and the direction are determined by applying the fuzzy membership function. A directed graph for representing the scenario in Fig. 3a is shown in Fig. 3b.

Further, we can represent other qualitative information about the nature of the terrain. This information is represented in terms of the attributes that are associated with the arcs in the graph. The values of an attribute can either be represented as exact quantities or can be partitioned into classes. We propose that these values be large partitions, so that the entire attribute can be described by a small number of values. For example, we restrict the values of the attribute curvature to be concave, convex or no curvature and that of the attribute inclination to be uphill, downhill or flat. The value of an attribute is defined with respect to a particular direction of traversal. Should the direction be reversed, the value of the the attribute (in general) will change. There is an intrinsic symmetry with this change in value. For example, traversing uphill in one direction will become downhill in the opposite direction. Alternatively, the attribute may have one value independent of the direction of traversal, for example, being slippery.

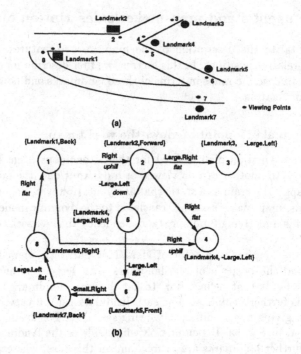

Fig. 3

To consider the reversal of an attribute when the direction of traversal is reversed, let the vector A represent the values of an attribute [a1, a2....an] and let F represent a $n \times n$ flip matrix defined as

$$
\begin{matrix}
0 & 0 & 0 & ...1 \\
0 & 0 & 0 & 1 & 0 \\
\\
1 & 0 & 0 & 0 & 0
\end{matrix}
$$

Then the reversal of the i-th value of the attribute A is given by the projection of the matrix product FA, that is ai-1 = PiFA The values of the attributes will play an important part when the control system for the agent is formulated at a later stage.

3.2 The Autonomous Agent's viewpoint: the navigation plan

From an agent's point of view, the actions performed can be perceived as events, and the current spatial and visual state of the agent can be represented as a situation.

Events are actions that the agent can perform. We distinguish two categories of events for a dynamic agent.

a) A visual event: this represents the act of noticing a landmark. It is a point event in time, and while it does not change the agent's spatial position, it alters his visual state and changes his knowledge about the environment.

b) A dynamic event: this represents the act of moving. It produces a change in spatial position, and this event has an attached attribute of direction.

A situation represents the current spatial and visual states of the agent. It consists of the components as shown in Table 1. First, the state of the agent being stationary or moving needs to be kept track of. An attribute of the situational aspect stationary is the spatial location, that is, at which viewing point is the agent positioned. If the situational aspect is moving, the associated attribute is direction, that is we record what direction the agent is moving in. A second aspect of the situation is the visual state. Here we distinguish two aspects: Expect_Landmark, which refers to the state of expecting to notice a landmark, and Notice_Landmark, which refers to the state of having recognized a landmark. In both these cases the associated attribute is a particular view of a landmark.

Aspects of a situation	Attribute
Stationary or Moving	If Stationary, at which viewing point
	If Moving, in what direction
Expect_Landmark or Notice_Landmark	Particular view of Landmark

Table 1

The process of navigating through the terrain can be seen in terms of a sequence of visual and dynamic events that change the situation. The objective of successful navigation is to go from an initial spatial situation to a final spatial situation, performing a sequence of events. This problem is a variation of the planning problem[1]. In the representation of a world map, since a node is a viewing point, to navigate, the agent must go from one viewing point to another. We use directed graphs to represent the agent's plan, in which the situations are nodes, and events are arcs. Consider traversing a terrain from viewing point 1, which is the back view of landmark 1, to viewing point 2, which is the front view of landmark 2. The initial situation describes the agent being stationary at viewing point 1, and noticing the back view of landmark 1 (see Fig. 4). Via Move_right the situation changes as the agent is in a state of motion headed in the right direction, and has an expectation of seeing the next visual event, which is the front view of landmark 2. The event See(Front View of Landmark 2) takes the agent to a new situation, where only the visual aspect changes from the previous situation. Move_right then takes the agent to the terminating viewpoint 2^1.

---

[1] We are not concerned here with the execution of these commands. For example, the first Move_right would require an agent to accelerate from zero velocity to some desired level, and the second Move_right would require the agent to maintain this velocity. The details can be found in [VeK91].

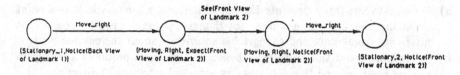

Fig. 4

3.3 Derivation of agent's navigation plan from the world map

One of the advantages of the world map we propose, is that this representation can be used for an automatic derivation of navigation plans for agents. An analysis of Figs 3(b) and 4 indicates the key elements of this automatic translation: going from viewing point 1 to 2 is represented by two nodes and an arc in the world map, and by four nodes and three arcs in the agent's plan.

To examine the process of automatically extracting the agent plan from the world map consider the the agent going from viewing point 1 to 2. The two nodes of the world map are the basis for the starting and the destination nodes of the agent's plan. The situation at the starting node is that of being stationary at landmark 1, and noticing the back view of landmark 1. The next two situations are added to reflect the agent anticipating and then noticing the next visual landmark. In this case, the next view is the front view of landmark 2. Finally the destination state reflects the agent being stationary at landmark 2.

The nature of visual events is determined by the destination state, i.e. if the destination state requires the agent to look at the front view of landmark 2, the visual states are Expect(Front View of landmark 2), and Notice(Front View of landmark 2), and the visual event that takes the agent from the situation of expecting a landmark to noticing a landmark is See(Front View of landmark 2).

The static events labelling arcs in the world map have to be altered to respective dynamic events that label the arcs in the agent plan that takes the agent from a stationary to a moving situation. For example, Right in Fig. 3b, will translate to Move_right in the navigation plan.

4 Discussion : Using the Navigation Plan

Although the emphasis of this paper is not on route planning, our formalism aids the process of route planning by taking the characteristics of the terrain at hand. Ignoring the characteristics of the terrain can lead to navigation plans that are unfeasible or expensive. Consider an autonomous agent be given the task of starting from 1, picking up an object from 5 and returning to 1. An heuristic could be used by the agent stating that when it is carrying something, an uphill path should be avoided. From Fig. 3a, the terrain along the route 1 - 2 - 5 involves going uphill, and the terrain along the route 1 - 8 - 7 - 6 - 5 is flat. The agent plans in an hierarchic fashion, in which the coarse level plan is first extracted, and then the finer details are filled in [VeK91]. At the coarse planning level the agent has a choice between a flat longer route, and a shorter uphill route. Initially, to pick up the object, the agent chooses the shorter route, since going uphill does not pose a problem. However, on the way back, while carrying an object, avoiding a hill is more important, and hence the longer route is selected. Each pair of nodes in the coarse level plan is then expanded according to the criteria in section 3.3.

4.1 Coping with Missing Landmarks

It is possible that in the process of navigation certain landmarks are obscured or not found. For example, to traverse from viewing point 1 to 2, the agent is expected to notice front view of landmark 2, and to go from viewing point 2 to 4, we notice the -(Large.Left) view of landmark 4. Should we restrict the agent to proceed from landmark 1 to 2 only if the anticipated visual event occurs, the agent will not successfully complete the traversal if the landmark is obscured. To make the system robust, we introduce the concept of current window of attention. The current window of attention defines a partial path extending from the agent's current position. For example, if the agent were at the viewing point 1, the current window defines the partial path 1 - 2 - 4. This window allows the agent to anticipate a sequence of visual events. For example, in the partial path 1 - 2 - 4, the sequence of visual events is:

Expect{Forward View of landmark 2}, Expect{-(Large.Left) View of landmark4}

The vision system expects this sequence. Thus in traversing from 1 to 2, if the agent does not encounter front view of landmark 2, but sees the -Large.Left view of landmark 4, the agent can reposition itself as being between 2 and 4. The agent can thus robustly traverse a given map, even if landmarks are missing or obscured.

5 Conclusions

We have proposed a new formalism for the representation of augmented maps. Crucial to this description is the use of visual landmarks. The directions contained within the map are expressed using a fuzzy set representation. The power of this representation lies in the fact that it is useful both as a static world map for the map maker, and can also be translated to a dynamic navigation plan for an autonomous agent.

The core of this representation is based on how humans give directions to travel from one place to another. This intrinsic property will enable extensions to our system whereby the autonomous agent will be capable of accepting directions directly from a person.

References

1. Georgeff,M.P.: Actions, Processes and Causality, in Reasoning about actions and plans. Proceedings of the 1986 workshop at Timberlin, Oregon, (M.P. Georgeff and A.L. Lansky, eds), Morgan Kaufmann, (1987) pp. 99-122.
2. Kender, J.R., Leff, A.: Why Direction Giving is Hard: The Complexity of using Landmarks in One Dimensional Navigation. AAAI-90 Workshop on Qualitative Vision, (1990) pp. 213 - 219.
3. Kuipers, B., Byun. Y-T.: A Robot Exploration and Mapping Strategy Based on a Semantic Hierarchy of Spatial Representation. AAAI'90 Workshop on Qualitative Vision, (1990) pp. 1 - 28.
4. Levitt, T.S. and Lawton D.T., 1990, Qualitative Navigation for Mobile Robots, Artificial Intelligence, vol. 44, pp. 305 - 360.
5. Levitt, T.S.: Qualitative Navigation. Proceedings of the DARPA Image Understanding Workshop, Los Angeles, Morgan Kaufmann (1987).
6. Stecter, L.A., Vitello, D. and Wonsiewicz, S.W.: How to tell people where to go. Comparing Navigation Aids, Journal of Man Machine Studies, vol. 22, (1985).
7. Tsuji, S., Zhang, J.Y.: The Qualitative Representation of Scenes along a Route, AAAI'90 Workshop on Qualitative Vision, (1990) pp. 67-71.
8. Venkatesh, S., Kieronska, D.: Landmark Based Representation for Maps and Navigation Plans. Tech. Report, Curtin University of Technology, (1991).
9. Zadeh,L.A., PRUF- a meaning representation language for natural languages, in Fuzzy Reasoning and its applications, Eds. Mamdani, E.H. and Gaines, B.R., pp. 1-66.

Using Fuzzy Logic in a Mobile Robot Path Controller

John Yen and Nathan Pfluger

Texas A&M University, College Station, TX 77840, USA

Abstract. A successful mobile robot system must have the ability to move about in a dynamic environment. Our approach to this problem is to plan a path for the robot to move from a starting point to a goal point using a map that may not have all the current information on the environment. The controller then dynamically reacts in an intelligent fashion to unplanned obstacles along the path by using sensor information and path information. This system attempts to follow a given path, adapting its course to the obstacles and shortcuts it perceives along the way. The key element of our approach is in using fuzzy logic to express and manipulate concepts needed for intelligent control. Some benefits from the use of fuzzy logic are ease of sensor fusion and generalization of desired direction of travel.

1 Introduction

The ability of a mobile robot system to plan and move intelligently in a dynamic system is needed if robots are to be useful in areas other than controlled environments. An example of a use for this system is to control an autonomous mobile robot in a space station, or other isolated area where it is hard or impossible for human life to exist for long periods of time (e.g. Mars). The system would allow the robot to be programmed to carry out the duties normally accomplished by a human being. Some of the duties that could be accomplished include operating instruments, transporting objects and maintenance of the environment.

There are many limitations of current approaches. Methods based on potential fields and stimulus-response paradigms have problems finding paths, even when they exist. The standard graph decomposition method always gives a path, but requires complete knowledge of the environment, and gives a path that is not easily followed. Finally, there are no approaches that have adequately addressed the problems involved with interleaving task planning, path generation and path execution.

The important issues that any realistic robot path planning system must have the ability to do are:

1. Plan several tasks concurrently.
2. Deal with a dynamic environment.
3. Deal with the problems of incomplete and/or inaccurate knowledge.
4. Work with the hindrance of limited sensing capability.

The main focus of our work has been on developing a fuzzy controller that will take a path and adapt it to a given environment. The robot only uses information gathered from the sensors, and avoids dynamically placed obstacles near and along the path. The problems of task planning and path generation will be dealt with in future research.

By using fuzzy logic, our project will be able to address the limitations of existing approaches. Our controller is able to use graph-decomposition and other high level path planning methods in a dynamic environment. Fuzzy logic techniques allow experts to express their planning and control rules in natural- language form. This makes the system easier to develop and more compact than standard logic systems.

2 Previous Work

In this section we will discuss previous approaches to mobile robot path planning, path execution, and path adaption. We will then give a background on the workings of fuzzy logic.

2.1 Mobile Robot Path Planning

The basic problem in mobile robot path planning is getting the robot from point A to point B without colliding into any walls or obstacles, in, hopefully, close to the shortest amount of time. Two basic methods have been proposed to accomplish this task.

The first method works by using a method called behaviors. The method is best compared to the biological paradigm of stimulus-response. The robot is given a start and a goal. The goal attracts the robot, while walls and obstacles repel the robot. Other behaviors, such as open-space finding, wall-hugging, or door-finding, can be added. This method of path planning, unfortunately, is sometimes unable to find paths, even though they exist. Another problem is that the robot sometimes becomes undirected, wandering about the map searching for doors and passages in an area where they do not exist.

The other basic method is based on graph decomposition. The space is partitioned into convex free areas, i.e. areas where the robot is allowed, which can be represented by a graph where each node is a free area and the presence of an edge between areas represents the fact that the areas touch. A graph search is then used to find a path from the start node to the goal node in the graph. Some major problems with graph-decomposition are that it gives a non-smooth path for the robot to follow and that all obstacles in the environment must be known and modeled for the robot to avoid them.

More recently, a new approach for controlling the robot was proposed by Payton[7] . This approach combined the two basic methods by giving the robot a non-smooth path and forcing it to avoid obstacles in the path. The approach was based on combining the output of the path following routine with data collected from the sensors. The robot would decide that it needed to turn by sensing an object in the forward sensor, and then would combine this knowledge with which way the path went to determine a final turn. The approach is similar to the one used for our controller.

2.2 Fuzzy Logic and Fuzzy Control

Fuzzy logic was first proposed by Zadeh[11, 12]. Fuzzy logic is based on the idea that humans do not think in terms of crisp numbers, but rather in terms of concepts. The degree of membership of an object in a concept may be partial, with an object being partially related with many concepts. By characterizing the idea of partial membership in concepts, fuzzy logic is better able to convert natural language control strategies which are used by humans to a form usable by machines.

Experience has shown[4, 5] that a controller based on fuzzy logic yields superior results than conventional control algorithms, and sometimes even better results than human operators. Fuzzy control appears to be most useful when the available sources of information are interpreted qualitatively, inexactly or uncertainly.

Some major advantages of fuzzy logic are that it allows a human expert to express his knowledge in a natural way, and that fewer rules in general are needed to express concepts, therefore saving time and space when a search for appropriate rules for a given situation must be conducted.

3 A Proposed Architecture

We now give an overview of the total system. Although most of the current research has been concentrated on the path execution subsystem, we wish to give an overview of a system where this subsystem will be useful. After defining our complete system, we will give the details involved in designing our path execution subsystem.

3.1 System Overview

Our architecture is based on the hierarchical approach to planning. This methodology allows us to divide the problem into levels of abstraction, inviting modular development. The current design of our architecture contains three principle layers.

1. Task Planner (TP): This planner works at the level where it must be decided which tasks or portion of tasks must be accomplished next. It must be able to coordinate multiple tasks that may or may not by independent. This planner decides the ordering of tasks, and determines the start point and goal point for each task. This planner works with the path generator to work out strategies for combining tasks.

2. Path Generator (PG): This planner uses a start point and a goal point given by the Task Planner and uses current maps to decide which is the best path to follow, based on the length and safety of a path.

3. Fuzzy Controller (FC): This subsystem, which is the actual controller for the robot, is given a path from a starting point to the goal point, and directs the robot along it. This controller works closely with the sensors to detect and avoid walls and obstacles, adapting the given path to the current situation. If the robot is forced too far from the path, then it has the option to request the higher level planners for a new plan.

3.2 Fuzzy Controller

Most of the early work has been done on the Fuzzy Controller. This controller is a natural extension of the controller proposed by Payton[7]. The largest difference is that where Payton uses discrete sets to represent the turn needed and the sensor readings, we use fuzzy sets and relations.

The controller is able to take a path, and follow it to the goal. The controller is able to cut corners on the path, while avoiding walls and obstacles, thereby adapting the path to the environment.

Overview The algorithm used by the Fuzzy Controller works in four steps.

1. Determine the angle between the current direction and desired direction along the path, irregardless of obstacles and wall. Fuzzify the result to make it more general.

2. Integrate the sensors using fuzzy sets to determine disallowed turning angles due to blocked sensors.

3. Combine the desired change in direction of movement and the disallowed change due to sensors.

4. Defuzzify the resultant combination to determine the control command for the change in direction.

Some sample paths and resultant execution that were given to the controller are shown in fig 1.

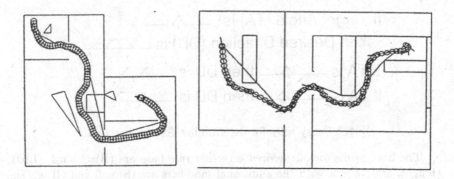

Fig. 1 Results of path execution.

The flow of data from the sensors and current path is shown in fig. 2. The shaded boxes represent areas where fuzzy sets and relations are used.

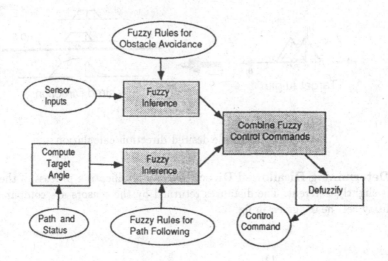

Fig. 2 Dataflow in the fuzzy controller.

Determining Desired Direction Fuzzification is performed in the first step by matching the crisp number corresponding to the target angle of travel to a fuzzy set representing the different types of turns. By computing the outcome of applying a set of fuzzy rules, for which a subset is given in fig. 3, the concept of wanting to turn by x degrees is broadened to the concept of turning in the general direction of x.

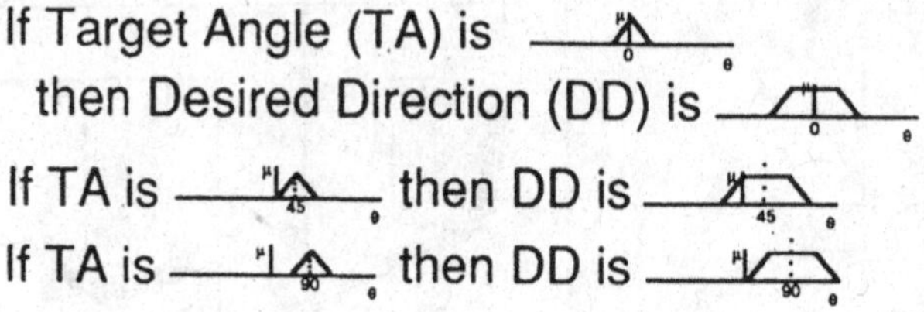

If Target Angle (TA) is _____ then Desired Direction (DD) is _____

If TA is _____ then DD is _____

If TA is _____ then DD is _____

Fig. 3 Fuzzy rules for determining desired direction.

The base terms for our desired direction rule base are (B)ackward, (L)eft, (R)ight, and (Z)ero, with the additional modifiers are (S)mall and (B)ig. The left hand side of the rule can be thought of as target angle, while the right hand side is the more general desired direction. This broadening of the target angle allows us to adapt the path to the environment by not constraining the robot to go only in direction of the target angle. An example calculation of a turn of thirty degrees is given in fig. 4.

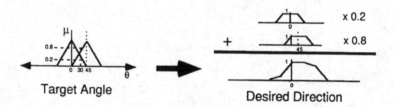

Fig. 4 Example desired direction calculation.

Determining Disallowed Direction The fuzzification process of the sensors is slightly different. The distances returned by the sensors are compared to the fuzzy set 'near', given in fig. 5.

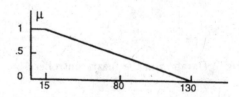

Fig. 5 Fuzzy set representing "new".

The degree of membership in near determines how disallowed that direction is. The fuzzy relations for determining the disallowed direction in three sensor directions is given in fig. 6. These sets are then combined using 'or' to determine a fuzzy set which expresses the idea of disallowed directions of travel for all angles. An example of combining sensors to determine disallowed direction is given in fig. 7, with the bold line indicating the final disallowed direction fuzzy set.

If -45° sensor distance to nearest object (SD) is *NEAR*

then Disallowed Direction (DD) is ⟋◺⟍

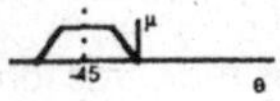

If 0° SD is *NEAR* then DD is ⟋◻⟍

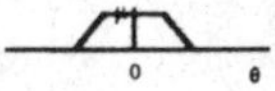

Fig. 6 Fuzzy rules for determining disallowed direction.

Combining disallowed direction directly with the desired direction does not make sense, in that one is positive information and the other is negative information. Therefore, we instead use the complement of disallowed direction, which is shown in fig. 8. This complement can be thought of as the allowed direction of the robot. This set combines directly with desired direction to obtain the final fuzzy control command.

Sensor Angle	Distance	μ_{near}
-90	20	0.9
-45	10	1.0
0	100	0.2
45	200	0.0
90	60	0.4

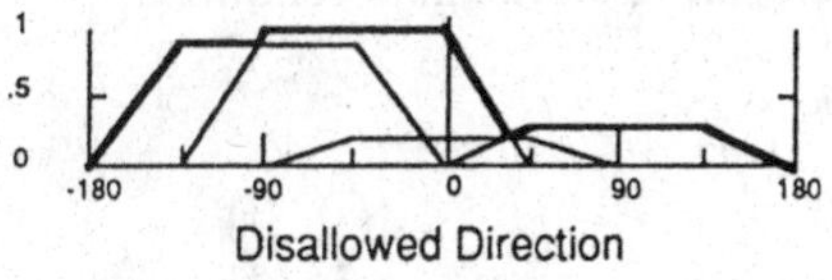

Disallowed Direction

Fig. 7 Example disallowed direction calculation.

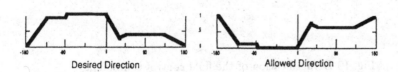

Desired Direction   Allowed Direction

Fig. 8 Obtaining the allowed direction.

Combining Desired Direction and Allowed Direction The process of combining the desired direction and the allowed direction is simple. In fuzzy logic terms the process is to take the fuzzy set that results as the minimum at each corresponding point between the desired direction and allowed direction. This process is shown in fig. 9. To obtain the final control command, the fuzzy set that results from the combination must be defuzzified.

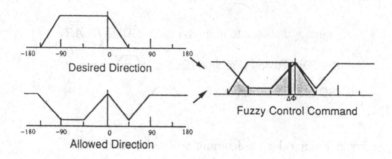

Fig. 9 Combining desired and allowed directions.

Defuzzifying the Result Defuzzification is needed to get a crisp number from our final fuzzy control command. There are three major defuzzification strategies, two of which are shown in fig. 10. The first, and simplest, strategy is to take the highest point along the set. This strategy has the drawback that only the strongest rule will affect the decision. We wish all parts of the fuzzy set to have some bearing on the final command.

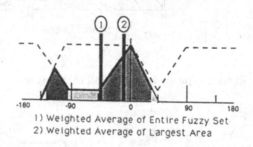

Fig. 10 Defuzzification of the final control command.

Therefore, the next strategy is to take a weighted sum of the set. This is shown as marker 1 in fig. 10. Although this strategy works well for normal fuzzy sets, it can lead to some problems, as shown. Since this set represents the combination of allowed and desired directions, we wish our final command to be in an area that is strong in both directions, and marker 1 is not strong in both.

Finally, we developed a strategy where we analyze the fuzzy set and break it into regions. We then take the weighted average of the largest region, shown as marker 2 in fig. 10. This technique satisfies both the conditions of using all (or most) of the set and giving a command that is usually strong in both desired and allowed.

3.3 Benefits

One of the major advantages of our control system is the robot's ability to cut corners and avoid unplanned obstacles. The desired angle is determined from a point that is well along the path, and this allows the robot to look ahead. In a situation where a corner is being approached, the robot knows both how to cut a corner, and when the turn is allowed by sensors. In addition, if an obstacle is directly in the path, the robot can avoid the obstacle while still traveling in the direction of the path.

4 Future Work

The next part of the path planner to be worked on is the Path Generator. By using aspects of the Fuzzy Controller, we hope to develop a simpler system than using graph-decomposition, which can become very computationally expensive for a large number of obstacles. Another problem that must be addressed is which path to take among the many possible paths that will be generated.

After the Path Generator is implemented, the task of implementing a real-time Task Planner is needed. The Task Planner will be in charge of scheduling tasks to be done by the Path Generator. The Task Planner must also deal with emergency situations, such as new high-priority orders or malfunctions with the robot. The methodology we are currently considering is using a scheduling algorithm that uses a fuzzy rule-based system to determine task priority for scheduling.

5 Conclusions

The goal of this research is to develop a fully functional robot in a dynamic environment. The current work being done on the control systems shows the high expressive power of fuzzy logic. It is hoped that the application of fuzzy logic to the planning portions of this system will be as fruitful. There however are still some problems to consider.

The first problem is how to determine the control and planning rules that a human expert uses. By mimicking a human operator, fuzzy control can usually do a more efficient job than the human it is mimicking due to the extra information available to the system. The next step is to determine what types of robot can be controlled by the Fuzzy Controller. Does the robot need to have vision or sonar sensors, or a combination of the two? How much deceleration and what maximum turning capacity can the controller handle? Is adapting the path to the given situation enough to handle all cases?

Finally, the rules for interconnecting the subsystems must be determined. When must the controller demand a new plan? What must the single task planner tell the multi-task planner? Hopefully the answers to these and other questions will be forthcoming in our research.

References

1. Albus,J. S.: A Theory of Intelligent Systems. In Fifth IEEE International Symposium on Intelligent Control, Philadelphia PA, (1990).
2. Anderson, T. L. Donath, M.: Synthesis of Reflexive Behavior for a Mobile Robot Based Upon a Stimulus-Response Paradigm. in: Mobile Robots III: Proceedings of the SPIE, V: 1007, Cambridge MA, (1988).
3. Arkin,R. C. : Motor Schema Based Navigation for a Mobile Robot: an Approach to Programming by Behavior. in: IEEE Conference on Robotics and Automation, pp. 164-271, (1987).
4. Lea, R.: Automated Space Vehicle Control and Rendezvous Proximity Operations. Telematics and Informatics, 5(3):179-185, (1988).
5. Lee, C. C. : Fuzzy Logic in Control Systems: Fuzzy Logic Controller - Part I. IEEE Trans. Systems, Man, and Cybernetics, 20:404-418, (1990).
6. Lee, C. C. : Fuzzy Logic in Control Systems: Fuzzy Logic Controller - Part II. IEEE Trans. Systems, Man, and Cybernetics, 20:419-435, (1990).
7. Payton, D. W. , Rosenblatt, J.K., Keirsey, D.M: Plan Guided Reaction. IEEE Trans. Systems, Man and Cybernetics, 20:1370-1382, (1990).
8. Pfluger, N., Yen, J., Langari, R.: A Defuzzification Strategy for a Fuzzy Logic Controller Employing Prohibitive Information in Command Formulation. in: Proceedings of the IEEE-FUZZ Conference, San Diego CA, (1992).
9. Yen, J., Pfluger, N.: Designing an Adaptive Path Execution System. in: Proceedings of the IEEE/SMC Conference, pp. 1459-1464, Charlottesville, VA, (1991) .
10. Yen, J., Pfluger, N.: Path Planning and Execution Using Fuzzy Logic. in: Proceedings of the AIAA Conference on Guidance, Navigation and Control, pp. 1691-1698, New Orleans, LA, (1991).
11. Zadeh, L. A.: Fuzzy Sets. Information Control, 8(3):338-353, (1965).
12. Zadeh, L. A.: Fuzzy Sets as a Basis for a Theory of Possibility. Fuzzy Sets and Systems, 1:3-28, (1978).

Part VI

Applications of Fuzzy Logic - Fuzzy Circuits

Fundamentals of Fuzzy Logical Circuits

Kaoru Hirota

Department of Instrument and Control Engineering, College of Engineering, Hosei University, 3-7-2 Kajino-cho, Koganei city, Tokyo 184, Japan

Abstract. Fuzzy logic is characterized as an extension of two valued Boolean logic. NOT, AND and OR operators in {0,1}-valued Boolean logic are extended to [0, 1]-valued fuzzy logic. They are called fuzzy negation, t-norm and s-norm (or t-conorm), respectively. These operators can be realized in electrical circuits.

A fuzzy logical circuit can be characterized as the integration of these fundamental circuits. The most important of these is the fuzzy inference circuit or fuzzy inference chip. Most inference circuits of this type are realized based on the min-max center of gravity method. This kind of fuzzy inference system is the foundation of industrial fuzzy applications, particularly in the field of fuzzy control. However, this technique is characterized as a fuzzy extension of combinatorial circuits in two valued Boolean logic. That is, such fuzzy inference schemas are repetitive although one stage inferences. There is therefore no need to think about memory modules or information transfer in the time axis.

In the case of AI applications, e.g. fuzzy expert systems, it is necessary to introduce multi-stage fuzzy inference. In such situations, the concept of a fuzzy extension of a sequential circuit, which is a complication of combinatorial circuits and memory modules in two valued Boolean logic should be discussed. Thus, it is essential to introduce the notion of fuzzy memory.

From this a point of view, the concept of fuzzy flip flop is presented in this paper. It is a fuzzy extension of a two valued J-K flip flop. The fundamental equations of several types of fuzzy flip flop are derived and their hardware implementations are shown. Finally, such fuzzy memory modules are combined with a fuzzy combinatorial circuit. The fundamental idea is presented in the context of the realization of fuzzy computer hardware.

1 Fuzzy logic

In classical {0,1}-two valued Boolean algebra, NOT, AND and OR operators constitute a complete and well-defined system of logic. Although there exists many variations in the concept of fuzziness, the unit interval [0, 1]-valued fuzzy concept is the most fundamental and widely accepted in practical applications. For this reason we consider only this definition in this article. If fuzzy logic (or [0, 1]-valued fuzzy logic) is characterized as an extension of the {0,1}-valued Boolean logic, it is necessary to investigate what the equivalent fuzzy operators of NOT, AND and OR would be. Before going into such a discussion, we would like to discuss another basic problem, i.e. how to express the $[0,1]$-value in a circuit.

Generally there are two methods to achieve this: either by analog or digital expressions. Of the analog expression, there are two further ways: the current mode expression and the voltage mode expression. For example, $[0\mu A, 5\mu A]$ and $[0V, 5V]$ are used in these expressions, respectively. In such expressions it is easier to understand their physical meaning, e.g. $3\mu A$ indicates the value 0.6 in the current mode expression, and $4V$ means the value 0.8 in the voltage mode expression. But the analog circuits are not so easy to implement compared with their digital counterparts when considering integration density, power consumption, heating problem, noise endurance and so on. For this reason, at present time, digital circuits are easy to use, where an approximation is necessary in the expression of a $[0,1]$-value. Usually 4 bit and sometimes 6 bit or 8 bit are used. In the 4 bit digital expression, the $[0,1]$-interval is approximately expressed by 16 states, i.e. $0000(=0), 0001(=1/15), \ldots, 1111(=1)$. If the bit parallel expression is used in the real implementation, the total number of bus lines causes a bottleneck problem. For this reason the bit serial expression is widely used, i.e. in the case of 4 bit expression one 4 bit value is transferred in the 4 fundamental clock periods on a single bus line.

Now let's move to the discussion of fundamental fuzzy logical operations. The fuzzy NOT operator is called the fuzzy negation and defined by the following axioms;

$$\textcircled{n} : [0,1] \longrightarrow [0,1] \tag{1}$$

$$n1 : 0^{\textcircled{n}} = 1 \tag{2}$$

$$n2 : a < b \to a^{\textcircled{n}} > b^{\textcircled{n}} \tag{3}$$

$$n3 : (a^{\textcircled{n}})^{\textcircled{n}} = a \tag{4}$$

These three axioms define the boundary conditions for NOT, the evaluation inversion, and the double negation, respectively. The most fundamental fuzzy negation is given by;

$$a^{\textcircled{n}} = 1 - a$$

But there exists infinitely many other fuzzy negation operations. Some of them are shown in fig. 1. In the case of real applications only (5) is used and some examples of such hardware circuits are shown in fig. 2. Fig. 2 (a) shows a 4 bit parallel digital fuzzy inverter, and it is easily confirmed that the bitwise inversion realizes (5). Fig. 2 (b) shows a 5V voltage mode analog fuzzy inverter circuit. It should be noted that this circuit is operated by clock pulses in the time domain. This is an important difference compared with the classical analog electric circuit.

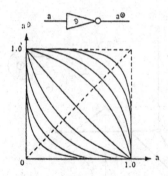

Fig. 1 various fuzzy negation operators.

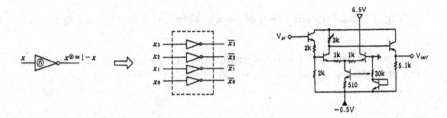

Fig. 2 examples of fuzzy inverter circuit.
(a) 4 bit parallel fuzzy inverter (b) 5V analog fuzzy inverter.

The fuzzy AND operator is called a <u>t-norm</u> (triangular norm) which is defined by

$$\textcircled{t} : [0,1] \times [0,1] \longrightarrow [0,1] \tag{5}$$

$$(a,b) \longrightarrow a \textcircled{t} b$$

$$t1 : a \textcircled{t} 0 = 0, a \textcircled{t} = a \tag{6}$$

$$t2 : a \textcircled{t} b = b \textcircled{t} a \tag{7}$$

$$t3 : a \textcircled{t} (b \textcircled{t} c) = (a \textcircled{t} b) \textcircled{t} c \tag{8}$$

$$t4 : a \leq b \rightarrow a \textcircled{t} c \leq b \textcircled{t} c \tag{9}$$

where the axiom $t1$: is a boundary condition including ordinary AND concept, $t2$: is the commutative law, $t3$: the associative law, and $t4$: the order preserving law. From a viewpoint of gate circuit implementation, $t2$: means that there is no need to distinguish input pin number 1 and input pin number 2 (i.e. symmetricity of two input pins), whereas $t3$: implies that the connection order is unimportant when 3 input (generally n input) t-norm gate is constructed by 2 input t-norm gates. Fig. 3 shows graphically the conditions obtained from $t1$: and $t2$:.

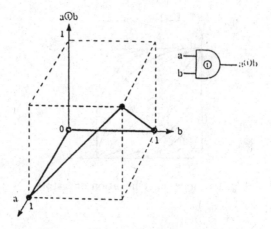

Fig. 3 boundary condition of t-norm obtained from $t1$: and $t2$:.

There are at least 50 t-norm operations, among which the most important and the most fundamental one is the logical product defined by;

$$a \ \textcircled{t} \ b = a \wedge b, \tag{10}$$

where $\wedge$ stands for a minimum operation. Other practically important t-norms are the algebraic product and the bounded product, and the drastic product which are defined;

$$a \ \textcircled{t} \ b = a \cdot b \tag{11}$$

$$a \ \textcircled{t} \ b = a \odot b = (a + b - 1) \vee 0 \tag{12}$$

where $\vee$ stands for the maximum operation, and

$$a \ \textcircled{t} \ b = a \wedge b = \begin{cases} b & a = 1 \\ a & b = 1 \\ 0 & \text{otherwise} \end{cases} \tag{13}$$

respectively.

Fig. 4 shows one example of logical product (or MIN) circuit in 5 volt analog mode.

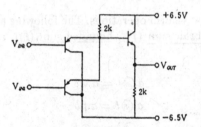

Fig. 4 Example of fuzzy AND gate in analog mode.

Finally the fuzzy OR operation will be discussed. It is called _s-norm_ (or _t-conorm_) and is defined in almost the same way as the t-norm, i.e.

$$\text{⑤} : [0,1] \times [0,1] \longrightarrow [0,1] \tag{14}$$

$$(a,b) \longrightarrow a \,\text{⑤}\, b$$

$$s1 : a \,\text{⑤}\, 0 = 0, a \,\text{⑤}\, 1 = a \tag{15}$$

$$s2 : a \,\text{⑤}\, b = b \,\text{⑤}\, a \tag{16}$$

$$s3 : a \,\text{⑤}\, (b \,\text{⑤}\, c) = (a \,\text{⑤}\, b) \,\text{⑤}\, c \tag{17}$$

$$s4 : a \leq b \rightarrow a \,\text{⑤}\, c \leq b \,\text{⑤}\, c \tag{18}$$

It is easy to see that the axioms are the same as those for the t-norm except for the boundary conditions $s1$ and $t1$. Fig. 5 illustrates the constraints of the s-norm induced by $s1 :$ and $s2 :$.

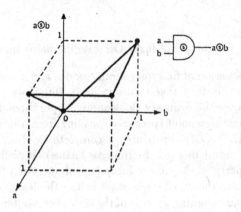

Fig. 5 the boundary condition of s-norm.

There are quite a few s-norm operations. The following are the most popular ones; <u>logical sum</u>, <u>algebraic sum</u> (†) , <u>bounded sum</u> ($\oplus$), and <u>drastic sum</u> (‡), which are defined by;

$$a \circledS b = a \vee b \tag{19}$$

$$a \circledS b = a \dagger b \tag{20}$$

$$a \circledS b = a \oplus b \tag{21}$$

$$a \circledS b = a \ddagger b = \begin{cases} b & a = 0, \\ a & b = 0, \\ 1 & \text{otherwise.} \end{cases} \tag{22}$$

respectively.

A typical example of a logical sum (or MAX) gate circuit in analog voltage mode is shown in fig. 6.

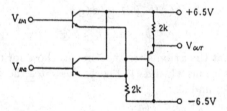

Fig. 6 Example of fuzzy OR gate in analog mode

In the above discussion of fuzzy negation, t-norm, and s-norm, a, b, and c are any fuzzy variables, i.e. they take values in $[0, 1]$. But if we restrict the values to $\{0, 1\}$, they become the ordinary Boolean logic operations NOT, AND, OR. From a mathematical viewpoint (precisely speaking from a lattice theory view), $(\{0, 1\}, \leq, NOT, AND, OR)$ constitutes a complete Boolean algebra (equivalently a complete complemented distributive lattice). It will be necessary to discuss such a property in the case of fuzzy logical operation $([0, 1], \circledN, \circledT, \circledS)$.

Firstly, de Morgan's law, which is valid in the Boolean logic, plays a very important role in the designing process of the digital computer circuit, so it may be natural to request the generalized version of de Morgan's law in the case of fuzzy logic, i.e.,

$$(a \circledT b)^{\circledN} = a^{\circledN} \circledS b^{\circledN} \tag{23}$$

$$(a \circledS b)^{\circledN} = a^{\circledN} \circledT b^{\circledN} \tag{24}$$

It is easy to confirm that each of $([0,1], \leq, 1 - \cdot, \wedge, \vee)$, $([0,1], \leq, 1 - \cdot, \dagger)$, $([0,1], \leq, 1 - \cdot, \odot, \oplus)$, $([0,1], \leq, 1 - \cdot, \wedge, \ddagger)$, satisfy the fuzzy de Morgan's law of (23) and (24), and they are called the <u>logical operation system</u> (or Zadeh operation system), <u>algebraic operation system</u> (or probabilistic operation system), <u>bounded operation system</u> (or Lukasiewicz operation system), and <u>drastic SAoperation system</u>, respectively. (It should be noted that one of the equations (23) and (24) can be derived from the other by using the axioms of fuzzy negation, namely t-norm, and s-norm.)

Among these 4 operation systems, the logical operation system (or Zadeh operation system) has the best property from a lattice theoretical viewpoint, i.e. it forms a complete pseudo Boolean algebra. It is thus easy to show that the system $([0,1], \leq, 1 - \cdot, \wedge, \vee)$, satisfies idempotent law, commutative law, associative law, absorption law, distributive law, double negation law, and de-Morgan's law. These properties are very convenient when we design fuzzy logical circuits or fuzzy controllers. This ease of hardware implementation may be one of the reasons why the Zadeh operation system (or MIN-MAX operation) is so widely used in fuzzy logic.

Finally, it should be noted that there is a big difference between the two valued Boolean logic and MIN-MAX fuzzy logic, i.e. the complemented law. This law holds true in two valued Boolean logic, i.e.

$$a \text{ AND } (\text{NOT } a) = 0, a \text{ OR } (\text{ NOT } a) = 1 \text{ for any } a = 0 \text{ or } 1 \qquad (25)$$

But in the MIN–MAX fuzzy logic,

$$a \vee (1 - a) \geq 0, a \vee (1 - a) \leq 1 \text{ for any } a \text{ from } \{0, 1\} \qquad (26)$$

2 Fuzzy Inference Chip

It will be natural to think about the integration of the fundamental fuzzy logical circuit described in section 2. A great deal of research has been directed toward the realization of a "fuzzy computer" and fuzzy inference chips have opened new opportunities for the field of intelligent control. Fuzzy inference engines are realized as microchips whose processors operate the rule based fuzzy inference with the operators we have already seen. The outline of the idea is summarized as follows.

In a conventional expert system, a human expert inputs specialized information by first expressing it in linguistic symbols. Afterwards, the system is capable of making reasoning decisions on the basis of this acquired operational knowledge. For instance, to take the simple example of controlling a hot room, the linguistic symbols might look like fig. 7.

```
IF (the room is) HOT
THEN (turn the air conditioner) ON.
(Now the room is) HOT.
(Turn the air conditioner) ON.
```

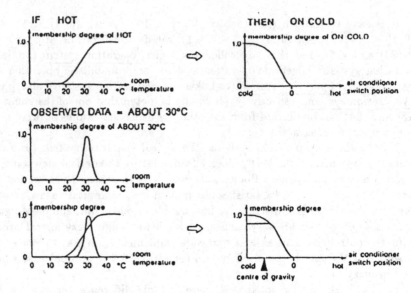

Fig. 7 An example of rule based fuzzy inference algorithm.

This can be interpreted that the machine is first given the knowledge that "HOT" leads to "ON", so that when its monitoring system tells it, "NOW the room is HOT", the system will draw the conclusion that the air conditioning should be turned on, and it will turn it on.

But how is "HOT" defined? Different people feel differently about the temperature in a room, and air conditioners can also be operated at various strengths. There is a continuum between "HOT" and "COLD", and between "HIGH" and "LOW". It would be prohibitively difficult to program a conventional expert system so sensitive that it could respond correctly to every point on such a continuum.

This is where the concept of the fuzzy set can be applied. The linguistic information is expressed in terms of membership of a fuzzy set. In the case of hot room, a temperature, say 30 degrees Celsius, is given a numerical value as a member of the fuzzy set "HOT". Then the knowledge "IF HOT THEN ON" can be collated with the information "Now 30C", and the air conditioner will be turned to "MEDIUM". As long as the numerical membership value of 30C in the concept "HOT" can be adequately expressed, the fuzzy inference machine can be constructed even to reflect personal preferences.

The fuzzy inference algorithm can express such individual differences, and at the same time it can improve the cost performance of such systems. There are a number of fuzzy inference algorithms, among which the so called MIN-MAX CG alqorithm seems to be a standard. Here CG stands for the center of gravity. The outline of this algorithm is shown in above fig. 7 (detailed explanations being omitted here).

Despite the emphasis of fuzzy in Japan, the world first fuzzy inference chip based on this algorithm appeared at AT&T Bell Laboratories in 1985, although the researchers responsible, [1], were Japanese. This was a CMOS 4 bit serial digital fuzzy chip and its architecture is shown in fig. 8.

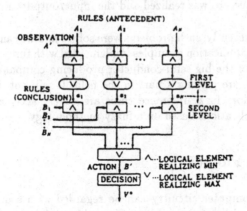

Fig. 8 Architecture of the world first fuzzy inference chip[1]

This fuzzy inference chip was designed on MULGA, which is a CMOS VLSI design software tool developed by Bell Lab., and consists of 2,300 transistors in ROM part and 6,000 transistors in inference part. The inference speed was 80,000 FLIPS(= Fuzzy Logical Inferences Per Second) and 1 chip covers 16 fuzzy rules. Both Togai and Watanabe retired from Bell Labs in 1986 and they continue their work at Togai Infra Logic Inc. and The University of North Carolina, respectively.

In 1988 Yamakawa[2] invented 5V voltage analog mode fuzzy inference chip, which is also based on MIN-MAX CG algorithm. In such a way several types of fuzzy inference chip were presented in various places. Almost all of them are hardware realizations of MIN-MAX CG algorithm. Many newspapers and magazines especially in Japan reported these fuzzy inference chips and some of them called the chip a "fuzzy computer", but the name was given too early because the technique was not, at that stage, mature.

Applications of fuzzy control became very popular in Japan. The first fuzzy wave appeared in the field of process control in Japan in 1987. The second also started in Japan in 1990, on this occassion, in the field of so called fuzzy home electronics. The total number of industrial applications now exceeds 300. Most such applications are based on the rule based fuzzy inference algorithm, mainly MIN-MAX CG method. However, most these applications (particularly the earlier ones) do not use fuzzy inference chips. Sixteen bit (nowadays thirty two bit) process computers were used in process control applications and the fuzzy inference algorithm has typically been implemented in software. In the case of fuzzy home electronics applications, a four bit microprocessor, plus a few KB ROM or RAM are used for address look up, and the fuzzy inference technique originally introduced by Mycom Inc. and the author's group[3], are mostly used.

There are several reasons why fuzzy inference chips were not used in the early applications. Some of them are as follows; (i) the cost was too prohibitive; (ii) the inference speed was too fast for the real application purposes; (iii) only the inference architecture was realized and the input/output interfaces were not supported.

However, more recently, such problems were solved and we can observe several concrete industrial application examples implemented with fuzzy inference chips. As a result, many of the big semi-conductor producing companies, e.g. Fujitsu, Oki, and Motorola, are playing a part in the field. It will not, in this authors' view, take long before fuzzy inference chips are widely accepted in industrial process control applications based on a fuzzy methodology.

3 Fuzzy Flip Flop (F^3)

Ordinary digital computer circuitry can be regarded as a sequential. The sequential circuit consists of the concept of combinatorial circuit and memory. Fuzzy logical circuits are an extension of ordinary digital computer circuitry. The fuzzy inference chip opened a new applicational field of fuzzy control. Despite this however all of them are based on single-step fuzzy inference and thus correspond to the extension of combinatorial circuits. To realize the extension of sequential circuitry, e.g. multistage fuzzy inference, fuzzy memory modules are indispensable. In the case of ordinary digital computers, a binary flip-flop circuit, which can memorize single bit of information, has been widely used as a fundamental element of memory modules. Such a binary flip-flop has been extended to a fuzzy flip-flop (F^3) by the authors' group[4]. The following is an outline of that circuit.

A flip-flop circuit, especially a J-K flip-flop that can memorize a single bit of information, has been of great use in memory modules of computer hardware. The next state $Q(t+1)$ of a J-K flip-flop is characterized as a function of both the present state $Q(t)$ and the present two inputs $J(t)$ and $K(t)$ (cf. tab. 1). A simplified notation J, K, and Q is sometimes used instead of $J(t), K(t)$, and $Q(t)$, respectively, in the following. The min-term expression of $Q(t+1)$ is

$$Q(t+1) = \overline{J}\,\overline{K}Q + J\overline{K}Q + J\overline{K}Q + JK\overline{Q} \tag{27}$$

simplified as

$$Q(T+1) = J\overline{Q} + \overline{K}Q \tag{28}$$

(where + in the right hand side means OR), which is well-known as a characteristic equation of a J-K flip-flop. On the other hand, another mutually equivalent max-term expression can be given and simplified as;

$$Q(t+1) = (J+Q) \cdot (\overline{K} + \overline{Q}). \tag{29}$$

$J(t)$	$K(t)$	$Q(t)$	$Q(t+1)$
0	0	0	0
0	0	1	1
0	1	0	0
0	1	1	0
1	0	0	1
1	0	1	1
1	1	0	1
1	1	1	0

Table 1. Truth table of J-K flip-flop

The equivalence of (26) and (27) can be easily confirmed using such well-known properties as double negation, de Morgan's law, commutative law, associative law, distributive law, absorption law, and complemented law, among which it should be noted that the last, not valid in fuzzy logic, is used.

Using fuzzy negation, t-norm, and s-norm, we can extend (26) and obtain;

$$Q_R(t+1) = (J \circledt Q^{\circledn}) \circleds (K^{\circledn} \circledt Q) \qquad (30)$$

In the same way (27) is extended to obtain

$$Q_S(t+1)v = (J \circleds Q) \circledt (K^{\circledn} \circleds Q^{\circledn}) \qquad (31)$$

It should be noted again that (26) and (27) are equivalent. In fuzzy logic however (28) does not always equal (29) because the complemented law and distributive law do not hold true. But it can be shown ([4]) that if the following relation concerning fuzzy negation, t-norm, and s-norm is valid;

$$A \circledt (B \circleds C) \geq (a \circledt B) \circleds (A \circledt C) \qquad (32)$$

then

$$Q_S(t+1) \geq Q_R(t+1) \qquad (33)$$

will be obtained.

Table 2 shows the values of (28) and (29) when J and K are restricted to two values $\{0,1\}$. If $J = 0, K = 1$ (i.e. if reset input is situated in the case of the J-K flip-flop), then the next state (28) is equal to zero (i.e. reset), but that of (29) is $Q \circledt Q^{\circledn}$. On the other hand, if $J = 1, K = 0$ (i.e. if set input is situated in the case of J-K flip-flop), then the next state (29) equals one, but that of (28) is $Q^{\circledn} \circleds Q (\leq 1)$. So we will define a reset-type fuzzy flip-flop by a characteristic equation (28), and a set-type fuzzy flip-flop by (29). (Here, note again that $Q \circledt Q^{\circledn}$ and $Q^{\circledn} \circleds Q$ are not always equal to zero and one, respectively, because of the lack of complemented law in fuzzy logic.)

To construct a fuzzy flip-flop circuit, we will discuss the fundamental system that is composed of $1-$ for fuzzy negation, min for t-norm, and max for s-norm.

In this system, (28) and (29) are expressed as;

J	K	Set Type (34)	Reset Type (33)	F^2
0	0	Q	Q	Q
0	1	$Q \; ⓘ \; Q^{(n)}$	0	0
1	0	1	$Q^{(n)} \; ⓢ \; Q$	1
1	1	$Q^{(n)}$	$Q^{(n)}$	$\bar{Q}$

Table 2 Values of (28) and (29) (J, K are restricted to $\{0, 1\}$)

$$Q_R(t+1) = \{J \wedge (1-Q)\} \vee \{(1-k) \wedge Q\} \qquad (34)$$

$$Q_S(t+1) = \{J \vee Q)\} \wedge \{(1-k) \vee (1-Q)\} \qquad (35)$$

respectively. A fuzzy flip-flop characterized by (32) is called a <u>min-max</u> reset-type fuzzy <u>flip-flop</u>, and that of (33) a <u>min-max</u> set-type <u>fuzzy flip-flop</u>.

In this case (31) holds true, but they are not equal. These two equations however can be united and the following is the chacteristic equation of <u>min-max</u> fuzzy <u>flip-flop</u>

$$Q(t+1) = \{J \vee (1-k)\} \wedge \{J \vee Q\} \wedge \{(1-k) \vee (1-Q)\}. \qquad (36)$$

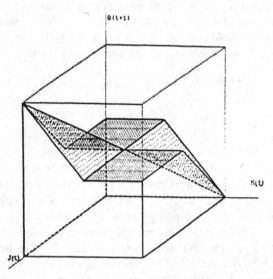

Fig. 9 Chacteristics of min-max fuzzy flip-flop when $Q(t) = 0.5$.

Electric circuits of this fuzzy flip-flop type have been implemented and tested. In addition, other types of fuzzy flip-flop have been studed and implemented. Foundations of designing fuzzy sequential circuits have also been reported[5].

4 Conclusion

As a basis of fuzzy logical circuit fuzzy inverter, t-norm gate, and s-norm gate are discussed first. Then fuzzy inference chips are described. Finally, the concept of fuzzy flip-flop was introduced. A fuzzy flip-flop should be considered the foundation stone from which to construct and design a fuzzy microprocessor.

References

1. Togai, M., Watanabe, H.: A VLSI implementation of fuzzy inference engine toward expert system on a chip. Proceedings of the 2nd IEEE International Conerence on AI and Applications,(1985), pp.192-191.
2. Yamakawa, T.,: Fuzzy Microprocessor-Rule Chip and Defuzzifier Chip. Proceedings of the International Workshop on Fuzzy System Applications, Iizuka Japan (1988), pp.51-52.
3. Arikawa, H., Hirota, K.: Address Look Up Virtual Paging Fuzzy Inference Chip. Journal of SICE, Vol.26, No.2, Feb (1990), pp.180-187 (written in Japanese)
4. Hirota, K., Ozawa, K.: The Concept of Fuzzy Flip-Flop. IEEE Trans on SMC, Vol.19, No.5, (1989), pp.980-997
5. Hirota, K., Pedrycz, W.: Designing sequential systems with fuzzy J-K flip-flops. International Journal of Fuzzy Sets and Systems, Vol.39, Feb.(1991), pp.261-278

Springer-Verlag
and the Environment

We at Springer-Verlag firmly believe that an international science publisher has a special obligation to the environment, and our corporate policies consistently reflect this conviction.

We also expect our business partners – paper mills, printers, packaging manufacturers, etc. – to commit themselves to using environmentally friendly materials and production processes.

The paper in this book is made from low- or no-chlorine pulp and is acid free, in conformance with international standards for paper permanency.

Lecture Notes in Artificial Intelligence (LNAI)

Lecture Notes in Computer Science